Nanoscale Electronic Devices and Their Applications

Nanoscale Electronic Devices and Their Applications

Khurshed Ahmad Shah
Farooq Ahmad Khanday

CRC Press is an imprint of the
Taylor & Francis Group, an **informa** business

First edition published 2021
by CRC Press
6000 Broken Sound Parkway NW, Suite 300, Boca Raton, FL 33487-2742

and by CRC Press
2 Park Square, Milton Park, Abingdon, Oxon, OX14 4RN

CRC Press is an imprint of Taylor & Francis Group, LLC

Library of Congress Cataloging-in-Publication Data
Names: Shah, Khurshed Ahmad, author. | Khanday, Farooq Ahmad, author.
Title: Nanoscale electronic devices and their applications / Khurshed Ahmad Shah, Farooq Ahmad Khanday.
Description: First edition. | Boca Raton : CRC Press, 2020. | Includes bibliographical references and index. |
Summary: "This book help readers to acquire a thorough understanding of the fundamentals of solids at nanoscale besides their applications including operation and properties of recent nanoscale devices. The book includes seven chapters covering overview of electrons in solids, carbon nanotube devices and their applications, doping techniques, construction and operation details of channel Engineered MOSFETs, structural and operational details about the spin devices including applications. Structural and operational details of phase change memory (PCM), memristor and Resistive Random-access Memory (ReRAM) are also discussed. Besides, some applications of these phase change devices to logic design have also been presented"—Provided by publisher.
Identifiers: LCCN 2020019398 (print) | LCCN 2020019399 (ebook) |
ISBN 9780367407070 (hardback) | ISBN 9780367808624 (ebook)
Subjects: LCSH: Nanoelectromechanical systems.
Classification: LCC TK7875 .S53 2020 (print) | LCC TK7875 (ebook) |
DDC 621.3815—dc23
LC record available at https://lccn.loc.gov/2020019398
LC ebook record available at https://lccn.loc.gov/2020019399

ISBN: 978-0-367-40707-0 (hbk)
ISBN: 978-0-367-80862-4 (ebk)

Typeset in Times
by codeMantra

Contents

Preface

Basic nanoscience research has been in existence for over 20 years. Since 2010, nanoscience and nanotechnology have made great strides in all fields of science and technology such as physics, chemistry, biology, engineering, and medicine. Nanoelectronics promise to be the foundation of the future industrial revolution. This technology deals with the characterization, manipulation, and fabrication of electronic devices at the nanoscale level with enhanced capabilities, reduced weight, and low power consumption. Electronic devices from computers to smart cell phones have become part and parcel of our life, and modern living is incomplete without these gadgets. However, increasing thermal issues and the manufacturing costs associated with traditional semiconductor technology have hindered further development in device technology. Nevertheless, nanoscale electronic devices are very small devices to overcome limits on scalability, which provide alternative options in terms of ease of processing, better flexibility, low cost, significant increase in speed, and more processing capability. Currently, for commercial use, devices with featured sizes of 14 nm or below can be fabricated, and new computer microprocessors have less than 50 nm of feature size. The superior electronic properties of materials come when electrons are confined to structures that are smaller than the distance between the mean free path of electrons in normal solids.

In the nanotechnology era, a variety of nanomaterials have been synthesized, which show unique physical, mechanical, electrical, electronic, and photonic properties. These nanomaterials have been used as functional elements in device applications; for example, carbon nanotubes, semiconductor nanowires, quantum dots, graphene, graphene oxide, and transition metal dichalcogenides have been used to replace rectifiers, junction transistors, field effect transistors, CMOS, RAM, and a number of other silicon-based devices. It is believed that nanoelectronic science will develop new nanoscale circuits, processors, and means of storing or transferring information, which can offer greater versatility because of faster data transfer, more on-the-go processing capabilities, and larger data memories.

While keeping in tune with developments in nanoelectronics, it is also important to see their applications in contemporary fields of technology. One such technology is Artificial Intelligence (AI). As of now, AI-based computing mostly deals with the software approach. However, some hardware implementations are also being perused. Therefore, it is important to comprehend the devices used for the implementation of AI systems. Given the complexity of the AI-based computing, nanoelectronic devices are going to greatly influence the achievement of the goals of AI-based computing.

Keeping in mind the futuristic impact of nanoscale electronic devices, it is imperative to discuss their advances and applications, and this book is an attempt in that direction. The inclusion of this text on advances in nanoscale devices in science curriculum will foster understanding of the subject at both undergraduate and graduate levels and will benefit scholars as well as students of science and engineering. The recent developments in the subject covered in this book will undoubtedly boost understanding and interest of students and researchers in fabrication of novel nanoscale electronic devices and their applications in the electronic industry and real life.

Written in an easily understandable style, the topics covered in this book are detailed enough to capture the interest of a curious reader and at the same time comprehensive enough to provide a necessary background to motivate further exploration of the subject. This book comprises seven chapters. All fundamental concepts that are very important to understanding the transport properties of nanoscale electronic devices, such as motion of electrons in solids, quantum transport, origin of band gap in solids, low-dimensional systems, calculation of their density of states, popular calculation methods for device transport, non-equilibrium Green's function (NEGF), and density function theory (DFT), are introduced in Chapter 1. Researchers and scientists across the globe during the past two decades have shown tremendous interest in the fabrication of nanoscale devices based on carbon nanotubes (CNTs). This is due to the fact that channel width of CNTs is on the order of 1 nm, much smaller than conventional silicon transistors and therefore more promising in higher device densities. Although they have low defect density, carrier distribution is not sensitive to temperature variations and has easier quantum confinement due to small channel width. Knowing the structural uniqueness of CNTs and its advantages in electronic device applications, Chapter 2 presents the structural parameters of CNTs and their extraordinary properties, in addition to its applications: two-probe devices, transistors, logic gates, sensors, photo-detectors, interconnects, and memories. It reveals that CNT-based devices have superior electronic characteristics than conventional semiconductor-based devices with a wide range of applications.

Chapter 3 presents various techniques of doping, transport properties of two-probe and three-probe devices, negative differential resistance (NDR) in CNTFETs, CNT-MOSFETs, and chromium-doped CNT devices, in addition to comparative studies of conventional and electrical doping. This chapter indicates that there is an advantage while doping CNT channels with various dopants and offers greater scope for various electronic device applications. Chapter 4 introduces the materials, which are going to dominate the nanoelectronic industry in the near future given the extraordinary features offered by them. These materials include graphene, transition metal dichalcogenides, and silicene/germanene. In addition, the reported FETs based on these materials have been introduced, and the challenges faced in the implementation of these FETs are also presented.

Chapter 5 introduces an overview of several non-linear and short-channel effects in MOSFETs and presents some non-conventional solutions to the problems arising due to miniaturization. These solutions include the employment of multiple gates, multiple materials for the gate(s), multiple doping regions in the channel, and the strained channel. In addition, the effect on several performance parameters of MOSFETs due to these solutions has been described. Finally, the performance analysis of a multi-gate, multi-material tunnel FET is given, which is considered to be an important device for future CMOS technology. In Chapter 6, first some of the basic concepts of spintronics, which is considered as an alternative for conventional charge-based electronics, are introduced, followed by the spin-based devices of magnetic tunnel junction (MTJ) and spin field-effect transistor (spin-FET) and their operation mechanisms. In addition, a brief overview of the challenges faced with the implementation of these devices is presented. Finally, the logic applications of these spin-controlled devices including non-volatile and reconfigurable logic design have been presented.

In Chapter 7, the theory, implementation, operation, and applications of the phase-change devices such as phase-change memory (PCM), memristor, and resistive random access memory (RRAM), which are considered as important devices for AI system implementation, are presented. In addition, a brief overview of the challenges faced with the implementation of these devices is presented.

Khurshed Ahmad Shah
Farooq Ahmad Khanday

Acknowledgments

We are greatly indebted to the many colleagues and friends who have contributed to *Nanoscale Electronic Devices and Their Applications* by commenting and making suggestions on selected chapters. The valuable contributions of Dr. Faisal Bashir, Dr. Javeed Iqbal Reshi, Muhammad Shunaid Parvaiz, Aadil Tahir, Mubashir Ahmad, Gul Feroz Ahmad Malik, Furqan Zahoor, Muzaffar Ahmad, Samrah Mehraj, Arshid Nisar, Zaid Mohammad Shah, Shazia Rashid, Hilal Ahmad Bhat, and many more are highly appreciated.

We are also thankful to our family members for their encouragement, support, and warmheartedness during the preparation of the manuscript of this book.

It has been a pleasure to work with Gagandeep Singh, Lakshay Gaba, Mouli Sharma, and Michele Dimont at CRC Press/Taylor & Francis Group, and we are truly grateful for all their professional help and positive spirit.

Last, but not the least, we are highly thankful to our parents, who have awakened our consciousness by equipping us with moral values, culture, and scientific temperament since our very childhoods and provided enough support and encouragement to allow us to complete this project in a fruitful manner. This book is lovingly dedicated to our parents.

Khurshed Ahmad Shah
Farooq Ahmad Khanday

Authors

Dr. Khurshed Ahmad Shah is currently working as senior assistant professor, Department of Physics, Sri Pratap College, Cluster University Srinagar, Jammu and Kashmir, India. He did his doctorate in physics from Jamia Millia Islamia, Central University, New Delhi, India and MS/MPhil degrees through University of Kashmir, Srinagar, India. He has published a good number of research papers in national and international refereed journals including IEEE, Elsevier, and Springer, and co-authored three books including *Nanotechnology: The Science of Small* with internationally reputed publisher John Wiley & Sons/Wiley, India. In addition, he has presented his research in many national and international conferences and guided research scholars. He has broad research interests in the areas of synthesis and characterization of 0D, 1D, and 2D materials and their applications, modeling, and simulation of nanoscale electronic devices. He is an editorial board member and reviewer of many scientific journals and member of many scientific and academic associations including Institute of Electrical and Electronics Engineers (IEEE) and International Association of Advanced Materials (IAAM).

Dr. Shah has successfully handled three major research projects as principal investigator, funded by University Grants Commission, Delhi, India, and Science and Engineering Research Board (SERB), Government of India. Under these projects, he has established two research laboratories. The main awards in his name include Indian National Science Academy (INSA) Visiting Scientist Fellowship (2019–2020), State Innovative Science Teacher Award 2013, Jawaharlal Memorial Fellowship for Doctorial Studies 2006, topped at the national level, Young Scientist Fellowship 2010, and three Jawaharlal Nehru Center for Advanced Scientific Research, Bangalore, India Visiting Scientist Fellowships.

Dr. Farooq Ahmad Khanday received BSc, MSc, MPhil, and PhD degrees from University of Kashmir in 2001, 2004, 2010, and 2013, respectively. From 2005–2009, he served as assistant professor in the Department of Electronics and Instrumentation Technology, University of Kashmir. In 2009, he joined the Department of Higher Education J&K and Department of Electronics and Vocational Studies, Islamia College of Science and Commerce, Srinagar, as assistant professor. In 2010, he joined as assistant professor in the Department of Electronics and Instrumentation Technology, University of Kashmir. His research interests include fractional-order circuits, nanoelectronics, low-voltage analog integrated circuit design, hardware neural network, quantum computing, stochastic computing, and biomedical circuit design. Dr. Khanday is the author or co-author of many publications in peer-reviewed, indexed international and national journals/conferences of repute and three book chapters. He is the Management Committee (MC) Observer of the COST Action CA15225 (fractional-order systems—analysis, synthesis, and their importance for future design) of the European Union. He is the senior member of IEEE and a member of several professional societies. He is serving as a reviewer for many reputed international and national scientific journals in electronics. He has successfully guided many PhD, MPhil, and an MTech theses. He also has completed/ ongoing funded research projects to his credit and has established laboratories with state-of-the-art facilities for pursuing research in the fields of IC design, nanoelectronics, fractional-order systems, etc.

1 Fundamentals of Nanoscale Electronic Devices

1.1 INTRODUCTION

Low-dimensional physics has revolutionized semiconductor device technology, as it relies on the technology of heterostructure, where the composition of a semiconductor can be changed on the scale of a nanometer [1]. By suitably controlling the properties of a nanoscale material, the futuristic devices could lead to beyond imagination device technology. Researchers across the world are studying many nanoscale devices, which could replace the existing silicon-based devices, as the end of Moore's law has arrived [2]. Carbon nanotube (CNT)-based field-effect transistors (FETs) and FinFETs are recognized as the most promising candidates for future electronic devices. The carbon nanotube field-effect transistor (CNT-FET) is regarded as an important contending device to replace silicon transistors, since CNTs are free of many existing problems associated with silicon technology [3]. For example, carrier transport is one-dimensional in CNTs, and the strong covalent bonding gives the CNTs high mechanical and thermal stability and resistance to electromigration [3]. Besides, the diameter of CNTs is controlled by its chemistry and not by the standard conventional fabrication processes [4]. In order to enhance the synthesis and functioning of the modern nanoscale electronic devices including those based on CNTs, graphene, silicene, germanene, molybdenum disulfide, and tungsten disulfide; channel-engineered devices; spin-controlled devices; and phase-change devices, the understanding behind their basic operating principles is of utmost importance. The functioning of these future nanoscale devices is controlled by a large number of parameters, which show a prominent physical significance [5].

In this chapter, we have explained the basic physics and features involved in the operation of these nanoscale electronic devices. The free electron model is successful in explaining many properties of metals, such as thermal conductivities, thermionic emission, and thermoelectric effect. However, this model fails to explain the properties of solids that are determined based on their internal structure. Quantum mechanics provides a clear picture of the nature at the subatomic scale and predicts all phenomena in terms of probabilities. The important thing to note here is that the mass of a nanomaterial is very small; therefore, the gravitational forces are negligible and electromagnetic forces are dominant in determining the behavior of atoms and molecules in these materials [6,7]. Furthermore, the quantum theory is used to explain their structure and properties, which contradict the results of the classical theory. Also, one important characteristic of materials that are being used for electronic devices is the origin of energy bands, as different solids possess different band structures, which give rise to a wide range of their electrical properties [8]. Depending on the nature of band occupation by electrons and the width of forbidden energy bands, all solids can be classified into conductors, semiconductors, and insulators. On the basis of these characteristics, the materials are used in different types of electronic devices for different applications.

For many electronic properties and general distribution of electrons in a material, the number of available states per unit volume per unit energy range needs to be determined [9]. The density of states (DOS) of a material depends upon its dimensionality, which is defined as the number of energy states at a particular level that electrons are allowed to occupy and is a measure of how closely the energy levels are present in the material [9,10]. Therefore, keeping in view its involvement in electronic transport properties of devices, we have explained and derived DOS in zero-dimensional (0D), one-dimensional (1D), two-dimensional (2D), and three-dimensional (3D) systems. The first principles approach based on non-equilibrium Green's function formalism (NEGF) and density functional theory (DFT) is explained as well. These calculations are being used frequently in the device simulations for obtaining various electrical transport properties of nanoscale electronic devices [11].

1.2 FREE ELECTRON THEORY AND QUANTUM THEORY

1.2.1 Free Electron Theory

At the end of the 20th century, Drude tried to explain the electronic structure of metals called free electron theory, which was later modified by Lorentz, has come to be known as Drude–Lorentz free electron theory of metals, and is based on the following assumptions:

1. The electrons move freely between the ions of metal crystal which constitute an electron gas.
2. There exists a strong electrostatic force between the positively charged ions and the negatively charged electron gas of a crystal.
3. The mutual repulsion between the electrons in a crystal is ignored.
4. The Maxwell–Boltzmann distribution law determines the velocity of the electrons that collide occasionally with the atoms of the crystal.

The free model is successful in explaining many properties of metals, such as thermal conductivities, thermionic emission, and thermoelectric effect; however, this model fails to explain the properties of solids, which are determined based on their internal structure, and also fails to explain why some solids are conductors and some insulators. Furthermore, this model gives an explanation why negatively charged electrons gets attracted towards the positive pole, constituting an electric current. In fact, the current depends on the electron–atom collision and is proportional to the applied voltage gradient, as per Ohm's law. In addition, this model explains excitation and de-excitation of electrons by the absorption and emission of light, respectively.

1.2.2 Quantum Theory

We know that the allowed energy levels of an electron bound to an atom are quantized. Here, the permissible energy levels are determined for a free electron restricted to remain within the crystal, but free to move within its confines. Let us first determine the restrictions imposed by the law of quantum mechanics on the energies of the electron inside the crystal. For mathematical simplicity, consider an electron limited to remain within a one-dimensional crystal of length "*L*." Assume that the potential energy within the crystal is constant and equal to zero; however, at the two ends of the crystal, the electron is prevented from leaving the crystal by a very high potential barrier (V_0 tends to infinity) as shown in Figure 1.1.

That is,

$$V(x) = \begin{cases} V_0 & \text{for } x \le 0 \text{ and } x \ge L \\ 0 & \text{for } 0 < x < L \end{cases} \tag{1.1}$$

Inside the crystal, the Schrodinger equation becomes

$$\frac{d^2\psi(x)}{dx^2} + \frac{8\pi^2}{h^2} mE\psi(x) = 0 \tag{1.2}$$

Any periodic potential can satisfy this equation, and for simple calculation, let us suppose that the general solution of Equation (1.2) is of the type

$$\psi(x) = A \sin kx + B \cos kx \tag{1.3}$$

where *A* and *B* are arbitrary constants to be determined by applying boundary conditions.

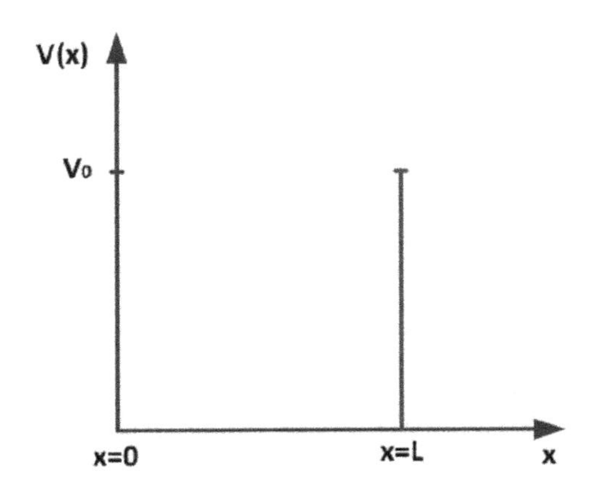

FIGURE 1.1 Illustration of a particle in a one-dimensional crystal of length *L*.

Since the electron is bound inside the crystal of length "L,"

$$\psi = 0 \text{ at } x = 0$$

$$\psi = 0 \text{ at } x = L$$

From Equation (1.3), we have $B = 0$,

$$\therefore \psi(x) = A \sin kx \tag{1.4}$$

using at $\psi = L$ at $x = L$, in the above equation, we get

$$\psi(x) = A \sin kL = 0$$

$$\sin kL = 0 = \sin n\pi$$

$$kL = n\pi$$

$$k = \frac{n\pi}{L}$$

where $n = 1,2,3,\ldots$ represents the order of the states.

If $n = 0$,

$$k = \frac{n\pi}{L} = 0$$

From Equation (1.4), we get

$$\psi(x) = A \sin 0 \Rightarrow \psi = 0$$

Therefore, $n = 0$ is not allowed.

From Equation (1.3), we have

$$\psi(x) = A \sin kx$$

$$\Rightarrow \psi_n(x) = A \sin \frac{n\pi}{L} x \tag{1.5}$$

For every value of n, there is a corresponding quantum state ψ_n, whose energy can be obtained from Equations (1.2) and (1.5) and is given by

$$E_n = \frac{h^2 k^2}{8\pi^2 m} = \frac{h^2}{8\pi^2 m}\left(\frac{n\pi}{L}\right)^2 = \frac{n^2 h^2}{8mL^2} \tag{1.6}$$

From Equation (1.6), it is clear that

1. The bound electrons can have only discrete energy values corresponding to $n = 1,2,3,\ldots$
2. The lowest energy of the particle is obtained if $n = 1$, such that

$$E_1 = \frac{h^2}{8mL^2}$$

$$\Rightarrow E_n = n^2 E_1$$

3. The spacing between the two consecutive levels increases as

$$(n+1)^2 E_1 - n^2 E_1 = (2n+1)E_1$$

The schematic representation of energy levels is shown in Figure 1.2.

The probability of the particle somewhere in the crystal must be equal to unity. That is,

$$\int_0^L P(x)\,dx = 1$$

$$\int_0^L |\psi_x|^2\,dx = 1$$

$$\text{or} \int_0^L A^2 \sin^2 \frac{n\pi}{L} x\,dx = 1$$

$$\Rightarrow A^2 \int_0^L \frac{1}{2}\left(1 - \cos 2\frac{n\pi}{L}x\right)dx = 1$$

$$\Rightarrow \frac{A^2}{2}\left[\int_0^L dx - \int_0^L \cos 2\frac{n\pi}{L} x\,dx\right] = 1$$

$$\text{or } \frac{A^2}{2}\left[x - \frac{L}{2\pi n}\sin\frac{2n\pi}{L}x\right]_0^L = 1$$

Since the second term of the equation is zero,

$$\frac{A^2}{2}[x]_0^L = 1$$

$$\text{or } \frac{A^2 L}{2} = 1$$

$$\Rightarrow A = \sqrt{\frac{2}{L}}$$

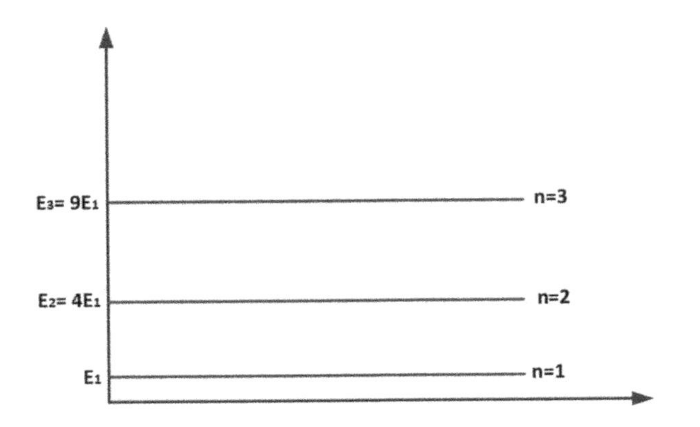

FIGURE 1.2 Representation of energy levels of a particle in a one-dimensional box.

Therefore, the normalized wave function (Equation 1.5) can be written as

$$\psi_n = \sqrt{\frac{2}{L}} \sin \frac{n\pi x}{L} \tag{1.7}$$

and the probability density is given by

$$|\psi_x|^2 = \frac{2}{L} \sin^2 \frac{n\pi x}{L}$$

All these results are contradictory to classical results:

1. Probability of finding the particle within the distance dx is the same anywhere in the box and is equal to dx/L and probability density = $1/L$ throughout the box, which is contrary to quantum mechanical results.
2. In classical mechanics, there is a continuous range of possible energies, whereas in quantum mechanics, the energy is quantized and so it cannot vary continuously, and hence the energy levels are discrete. However, if the particle becomes heavier and the length of crystal is large, the energy levels will be spaced very closely together and may become continuous in a gradual manner. Furthermore, the potential barrier confining an electron in the interior of an actual crystal is infinitely high and is determined in a complex way by the surface energies of the crystal. If this potential barrier at the surface of the crystal is high but not infinite, the wave function has the form as shown in Figure 1.3. The wave function is sinusoidal in the region $0 \le x \le L$ and exponential outside this region. It is expected that the extension of the wave function beyond the potential barrier is inversely proportional to the height of the barrier. Furthermore, if the barrier is narrow, it is possible that the wave function can extend beyond it. In this case, there is a finite but small probability $|\psi|^2$ of finding the electron on the other side

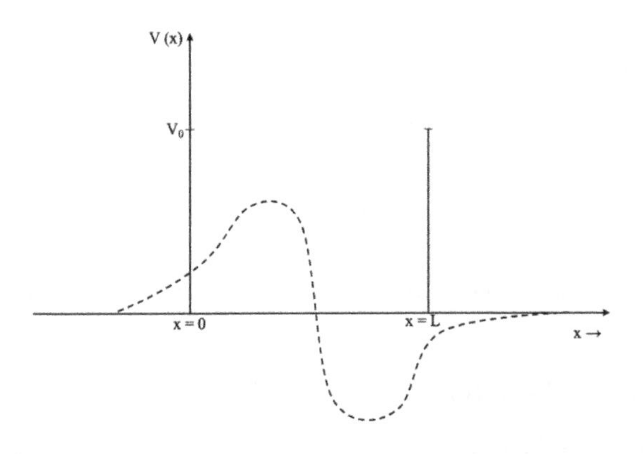

FIGURE 1.3 Representation of the wave function of the particle when potential barrier at the surface of the crystal is high but not infinite.

of the barrier. The ability of an electron to tunnel to a potential barrier is called the tunneling effect, which is a direct consequence of the application of quantum mechanics to this problem.

3. In three dimensions, the crystal can be approximated by a cube of side "L", inside which the potential energy is zero and outside which it is very high ($\rightarrow\infty$). Under this assumption, the solution to the three-dimensional Schrodinger equation is given by

$$\psi(x,y,z) = A_x \sin k_x x\, A_y \sin k_y y\, A_z \sin k_z z$$

where $k_x = \frac{n_x\pi}{L}$, $k_y = \frac{n_y\pi}{L}$, $k_z = \frac{n_z\pi}{L}$, and n_x, n_y, and n_z are integers >0.

Under the boundary conditions,

$$x = 0, x = L, y = 0, y = L, z = 0, z = L, \text{we get}$$

$$A_x = \left(\frac{2}{L}\right)^{\frac{1}{2}}, A_y = \left(\frac{2}{L}\right)^{\frac{1}{2}}, A_z = \left(\frac{2}{L}\right)^{\frac{1}{2}}$$

Therefore, the normalized wave equation becomes

$$\psi_n = \left(\frac{2}{L}\right)^{\frac{3}{2}} \sin\frac{n_x\pi}{L}x \sin\frac{n_y\pi}{L}y \sin\frac{n_z\pi}{L}z$$

The corresponding form of energy is given by

$$E_{n_x n_y n_z} = \frac{h^2}{8mL^2}\left(n_x^2 + n_y^2 + n_z^2\right)$$

Further,

$$\frac{P^2}{2m} = \frac{h^2}{8mL^2}\left(n_x^2 + n_y^2 + n_z^2\right)$$

$$\text{or} \left(\frac{2L}{h}\right)^2 P^2 = \left(n_x^2 + n_y^2 + n_z^2\right)$$

Hence, the momentum of an electron can be expressed directly by its three quantum numbers.

1.3 ORIGIN OF BANDGAP IN SOLIDS

Electrons in an isolated atom possess discrete energy levels (1s, 2s, 2p, etc.). These energy levels are filled with electrons in order of increasing energy. When these isolated atoms combine to form a solid, they arrange themselves in an ordered pattern, called a crystal. In a crystal, each atom is in the electrostatic field of neighboring atoms due to periodicity. The discrete energy levels of individual atoms are no longer valid. Due to interaction between the atoms, each discrete level splits into closely spaced sublevels. The number of sublevels is equal to the number of atoms N in the solid, $N \approx 10^{23}/\text{cm}^3$. Therefore, separation between the sublevels is very small (10^{-23}eV). Hence, these sublevels are almost continuous in energy and thus form energy bands. The first energy levels of various atoms form first energy band, the second energy levels form the second energy band, and so on. The energy band formed by valence electrons of atoms is called valence band. This band is the highest occupied band. The next higher band is known as conduction band and is normally empty. These allowed energy bands are in general separated by regions which have no allowed energy states; such regions are termed as energy gaps or forbidden energy bands.

The splitting of energy levels does not take place from lower energy levels 1s and 2s because the electrons in these levels are not significantly affected by the presence of other atoms. Further, the 2p level does not begin to split into sublevels until the interatomic separation becomes smaller than actually found as in the case of sodium. In fact, 3s level is the first occupied level to be split into sublevels. In higher energy levels, splitting occurs because electronic wave function overlaps significantly to give rise to the interaction between them. Different solids possess different band structures, which give rise to a wide range of electrical properties observed in various materials. Depending on the nature of band occupation by electrons and the width of forbidden energy bands, all solids can be classified into conductors, semiconductors, and insulators.

1.3.1 NEARLY FREE ELECTRON MODEL

In this model, the crystal potential is assumed to be very weak as compared to electron kinetic energy, so that electrons behave essentially like free particles. The weak periodic potential introduces only a small amount of perturbation effect on the free electron in the solid. However, the periodicity cannot be ignored. The allowed energy states are no more confined to a single parabola in the K-space, but states are represented by other parabolas as well, that are displaced by any reciprocal lattice vector $\vec{G}$.

$$\therefore E(k) = E(k+G) = \frac{\hbar^2}{2m}|k+G|^2$$

On the grounds of simplicity, parabolas for 1D crystal are shown in Figure 1.4.

The periodicity in the real space is "a," and in the k-space, it is given by the reciprocal lattice vector:

$$\vec{G} = \frac{n2\pi}{a}$$

At zone boundaries, the values are degenerate as the two parabolas intersect here. The first zone boundary occurs at $k = \pm\pi/a$. Therefore, the electron wave functions with these k values must be represented by a superposition of plane waves, which for a small potential can be taken as $\exp.\left(\frac{i\vec{G}.\vec{x}}{2}\right)$ and $\exp.\left(\frac{-i\vec{G}.\vec{x}}{2}\right)$. These waves move in opposite directions. In all these special values of $k = \pm n\pi/a$, the wave functions are made up of equal parts of wave traveling to the right and the left.

The other reciprocal vectors can be ignored in the approximation for the conduction of zone boundary. The wave functions may be expressed as

$$\left.\begin{aligned} \psi_+ &\approx e^{\frac{i\vec{G}.\vec{x}}{2}} + e^{\frac{-i\vec{G}.\vec{x}}{2}} \approx \cos\frac{\pi x}{a} \\ \psi_- &\approx e^{\frac{i\vec{G}.\vec{x}}{2}} - e^{\frac{-i\vec{G}.\vec{x}}{2}} \approx \sin\frac{\pi x}{a} \end{aligned}\right\} \tag{1.8}$$

These standing waves appear as a result of Bragg's reflections occurring at $k = \pm\pi/a$ or $\pm G/2$. Probability densities of the two sets of standing waves are

$$\left.\begin{aligned} \psi^*_+ \psi_+ &= |\psi_+|^2 \approx \cos^2\frac{\pi x}{a} \\ \psi^*_- \psi_- &= |\psi_-|^2 \approx \sin^2\frac{\pi x}{a} \end{aligned}\right\} \tag{1.9}$$

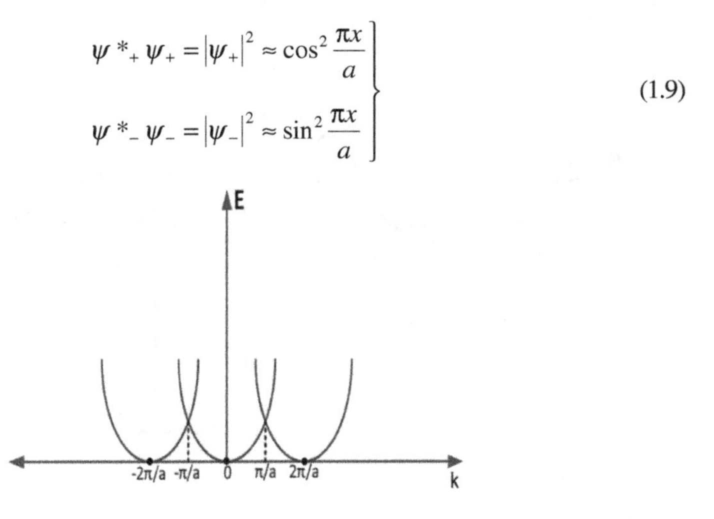

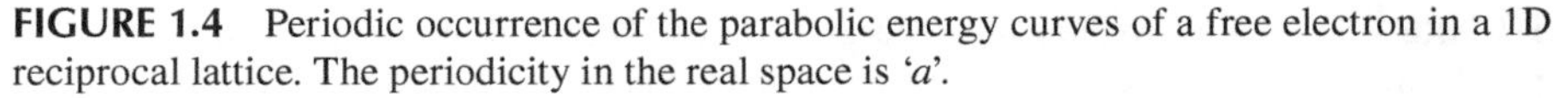

FIGURE 1.4 Periodic occurrence of the parabolic energy curves of a free electron in a 1D reciprocal lattice. The periodicity in the real space is 'a'.

The electron potential energy in 1D crystal is shown in Figure 1.5a. The potential field belongs to the positive ions (cores) whose valence electron moves in the field. The probability distribution of the standing waves and simple plane waves are shown in Figure 1.5b.

The plane waves exp.$(i\vec{k}.\vec{x})$ have the same probability density at all points, since exp.$(i\vec{k}.\vec{x})$.exp$(-i\vec{k}.\vec{x}) = 1$. The distribution of electron of ψ_+ favours in the piling of electron charges exactly above the cores. However, ψ_- pushes the electron charge away from the ion core.

Calculation of expectation values of potential energy over these charge distributions shows that the potential energy of $|\psi_+|^2$ is lower than that of the traveling wave, whereas the potential energy of $|\psi_-|^2$ is higher than that of the traveling wave. The energy gap of width E_g occurs due to the difference in potential energies

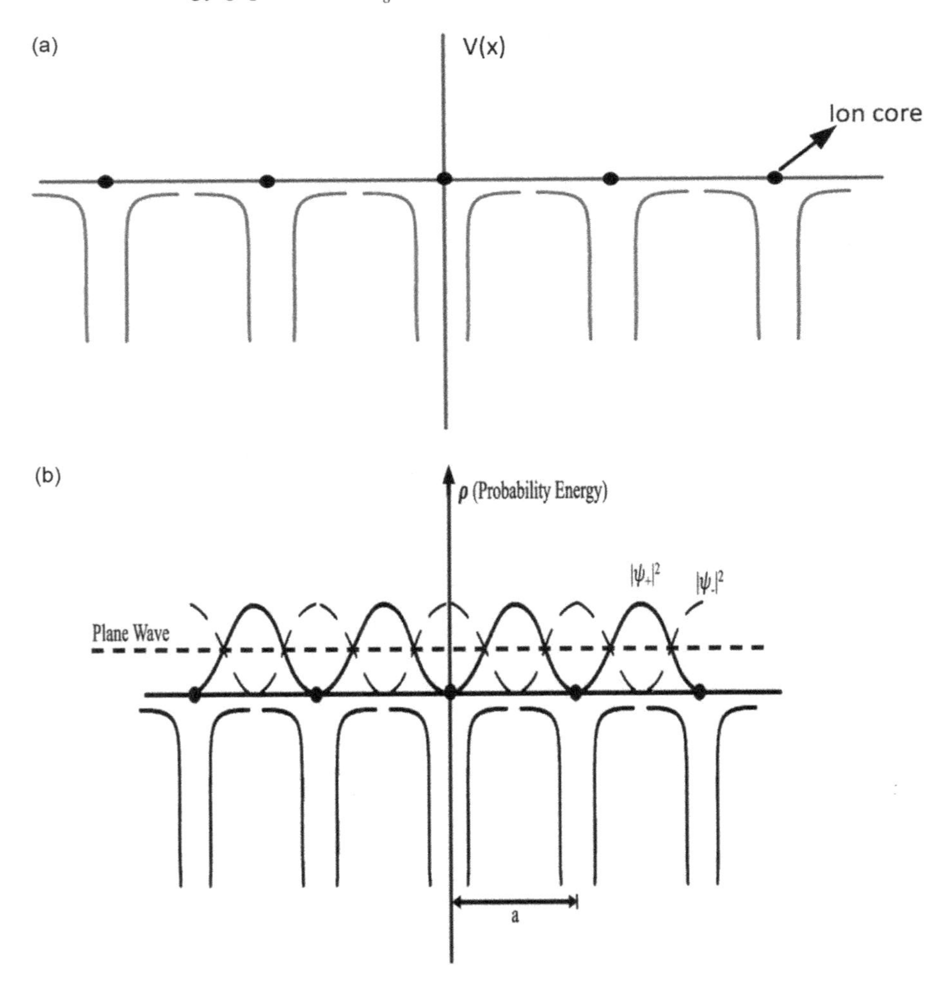

FIGURE 1.5 (a) Electron potential energy $V(x)$ in a 1D crystalline solid. (b) Distribution of probability density for the standing waves $\psi_\pm$ and the plane wave inside the crystal.

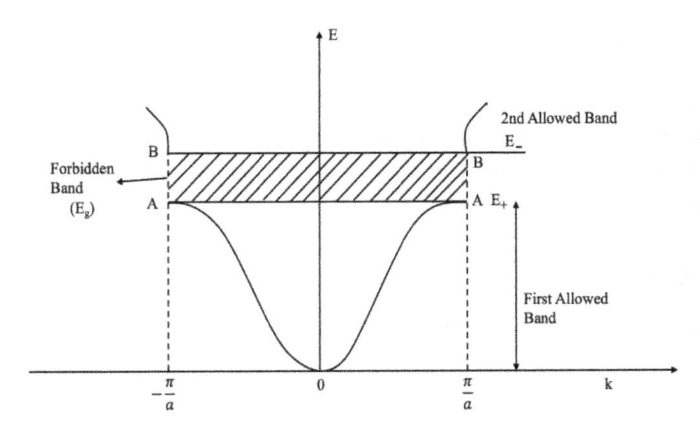

FIGURE 1.6 Plot of E versus k for an electron in a monoatomic linear lattice of lattice constant "a," the energy gap (E_g) associated with the first Bragg reflection at $x = \pm n\frac{\pi}{a}$ $(n \neq 0)$ where "n" is an integer.

corresponding to $|\psi_+|^2$ and $|\psi_-|^2$, which is shown in Figure 1.6. This is the region of band gaps observed in the energy band structure.

We can write the potential energy of an electron as a weak harmonic potential as

$$U(x) = 2U_1 \cos\frac{2\pi}{a}x \tag{1.10}$$

where U_1 is the Fourier coefficient.

The first-order energy difference between the standing wave states is

$$E_g = \Delta E = E_+ - E_- = \int_0^L U(x)\left[|\psi_+|^2 - |\psi_-|^2\right]dx \tag{1.11}$$

Normalizing the wave functions at $k = \pm\pi/a$ over the crystal length L, we get

$$\left.\begin{aligned} \psi_+ &\approx \left(\frac{2}{L}\right)^{\frac{1}{2}} \cos\frac{\pi x}{a} \\ \psi_- &\approx \left(\frac{2}{L}\right)^{\frac{1}{2}} \sin\frac{\pi x}{a} \end{aligned}\right\} \tag{1.12}$$

Using Equations (1.10) and (1.12) in (1.11), we get

$$\Delta E = \frac{2U}{L}\int_0^L \left(1 + \cos\frac{4\pi}{a}x\right)dx$$

$$\text{or } E_g = \Delta E = 2U$$

Thus, the band gap is equal to twice the magnitude of the leading Fourier coefficient of the crystal potential.

The eigenvalue of ψ_+ is lower in energy since the maxima of its probability density occur at the points of minimum potential energy. The plane wave energy at the zone edge is centered between two eigenvalues E_+ and E_- corresponding to ψ_+ and ψ_- in that order.

1.3.2 Approximate Measure of Band Gap

The Fourier expansion of one-dimensional crystal potential has the following form:

$$V(x) = \sum_{\vec{G}} V_{\vec{G}} e^{i\vec{G}.\vec{x}} \tag{1.13}$$

Taking into account that the potential energy function is real, Equation (1.13) can be rewritten, considering only the shortest reciprocal lattice vector as

$$V(x) = 2v\cos\frac{2\pi x}{a} \quad \text{with} \quad v = |V_{\vec{G}}| = |-V_{\vec{G}}| \tag{1.14}$$

Since the magnitude of $V_{\vec{G}}$ decreases as $\vec{G}$ increases. For an approximate calculation, we can ignore the contribution from the larger reciprocal vectors.

Using the first-order perturbation theory, the band gap is written as

$$E_g = \Delta E = E_+ - E_- = 2v\int \cos\frac{2\pi x}{a}\left[|\psi_+|^2 - |\psi_-|^2\right]dx \tag{1.15}$$

Normalizing the wave function at $k = \pm\pi/a$ over the crystal length L, we have

$$\left.\begin{aligned} \psi_+ &\approx \left(\frac{2}{L}\right)^{\frac{1}{2}} \cos\frac{\pi x}{a} \\ \psi_- &\approx \left(\frac{2}{L}\right)^{\frac{1}{2}} \sin\frac{\pi x}{a} \end{aligned}\right\} \tag{1.16}$$

Using Equation (1.16), we have, from Equation (1.15),

$$\Delta E = \frac{2v}{L}\int_0^L \left(1 + \cos\frac{4\pi}{a}x\right)dx$$

$$\Delta E = 2v$$

Thus, the band gap is equal to twice the magnitude of Fourier coefficients of the crystal potential. The range of allowed energy states covered by the dispersion curve is shown in Figure 1.6. The first Brillouin zone constitutes the first energy band.

1.3.3 Effective Mass Approximation

For a free electron, the energy depends on the wave vector given by

$$E = \frac{\hbar^2 k^2}{2m} = \frac{p^2}{2m}$$

where $\vec{p} = \hbar\vec{k}$ represents true momentum of the electron. However for an electron moving in a periodic potential, $\hbar\vec{k}$ does not represent the true momentum. This is because the energy does not vary as it does for a free electron. In a crystal, the momentum associated with an electron is called the crystal momentum. It is this quantity with respect to which we study the dynamical behavior of the electrons in a periodic potential. We use the conservation of crystal momentum and not of the momentum. On similar grounds, the mass of an electron in a crystal is not its true mass, but this is called effective mass of an electron.

Consider an electron initially in a state K, acted upon by external field E. In order to avoid any complication due to Pauli's exclusion principle, we assume that this is the only electron in the Brillouin zone. When the electric field E acts on the electron for a small time "dt," suppose that it has gained a velocity "v" over a distance "dx" during the time, then the energy is given by

$$d\varepsilon = -eEdx = -eEvdt \tag{1.17}$$

Where

$$v = \frac{1}{\hbar}\left(\frac{d\varepsilon}{dk}\right)$$

Therefore, Equation (1.17) can be written as

$$d\varepsilon = \frac{-eE}{\hbar}\left(\frac{d\varepsilon}{dk}\right)dt$$

or

$$d\vec{k} = \frac{-e\vec{E}}{\hbar}dt$$

$$\hbar\frac{d\vec{k}}{dt} = -e\vec{E}$$

or

$$\frac{d\vec{P}}{dt} = -e\vec{E} \tag{1.18}$$

Now,

$$\frac{dv}{dt} = \frac{1}{\hbar}\frac{d^2\varepsilon}{dk^2}\frac{dk}{dt} \tag{1.19}$$

Substituting Equations (1.18) in (1.19), we get

$$\frac{dv}{dt} = \frac{-eE}{\hbar^2}\frac{d^2\varepsilon}{dk^2}$$

or

$$-eE = \frac{\hbar^2}{\left(\frac{d^2\varepsilon}{dk^2}\right)}\frac{dv}{dt}$$

Comparing the above equation with Newton's second law, we have

$$m^* = \frac{\hbar^2}{\left(\frac{d^2\varepsilon}{dk^2}\right)} \tag{1.20}$$

If an electron moves in a uniform potential, then

$$\varepsilon = \frac{\hbar^2 k^2}{2m}$$

or

$$\frac{d^2\varepsilon}{dk^2} = \frac{\hbar^2}{m}$$

Therefore, from Equation (1.20), we get

$$m^* = m$$

But for an electron moving in a periodic potential, ε does not vary with k as it varies for a free electron. Thus, $m^* \neq m$. The effective mass of the electron in a crystal varies in a complex manner because of the complex nature of ε versus k curves. The electron in this curve will thus behave dynamically just like a particle with variable mass, and the whole effect of periodic potential on the motion of electron is to replace the free electron mass with proper effective mass. The significance of this is that the free electron theory of metal is justified to a great extent, provided we use effective mass instead of true mass. This is known as effective mass approximation.

1.4 TIGHT BINDING APPROXIMATION

A tight binding is one of the ways of evaluating energy levels of an electron in a solid. For materials which are formed from closed-shell atoms or ions, or even covalent solids, the free electron model seems inappropriate. In the tight-binding model, we imagine how the wavefunctions of atoms or ions will interact as we bring them together. Here, one starts with the wave function of an electron in a free atom and then constructs a crystal orbit, i.e., the Bloch function which describes the electron behavior in the periodic field of the crystal as a whole. Here, the single electron wave function in the crystal is expressed as a linear combination of the atomic orbitals (LCAO) that the electron occupies in a free atom. As a result, the discrete energy levels of a free atom will broaden into energy bands as the atoms are brought closer in the crystal. Moreover, this approximation is valid only for electrons corresponding to the inner electronic shell in the atom. It has successfully been applied to the d-electrons in transition metals and to the valance electrons in diamond-like and inert gas crystals. The potential energy of an electron in the field corresponding to the nucleus and that of the solid is shown in Figure 1.7.

Consider an S-electron in an isolated atom positioned at $\vec{r}_n$ with the ground state wave function $\phi(\vec{r}-\vec{r}_n)$, where r determines the electron position in space as shown in Figure 1.8.

The one-electron Schrodinger wave equation for free atom is given by

$$H_0\phi(\vec{r}-\vec{r}_n)=\sum E_0\phi(\vec{r}-\vec{r}_n) \tag{1.21}$$

where H_0 and E_0 are the Hamiltonian and the ground energy of the electron in a free atom, respectively.

The Hamiltonian of an electron in the crystal is expressed as

$$H = H_0 + V(\vec{r}-\vec{r}_n) \tag{1.22}$$

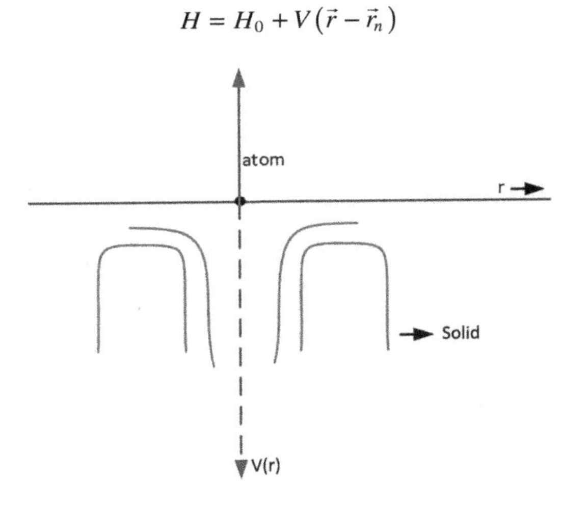

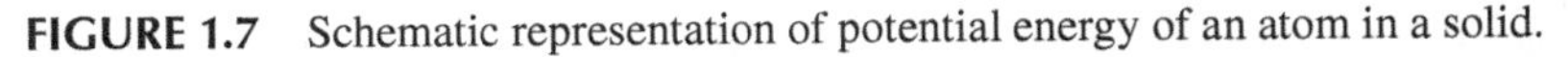

FIGURE 1.7 Schematic representation of potential energy of an atom in a solid.

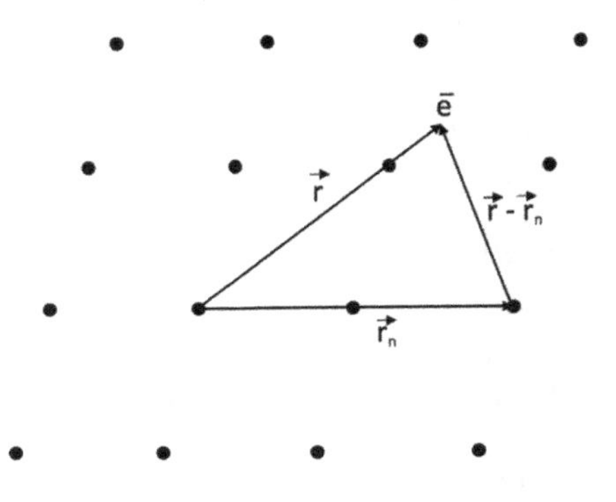

FIGURE 1.8 An S-electron in an isolated atom positioned at $\vec{r}_n$.

where $H_0 = \frac{-\hbar^2}{2m}\nabla^2 + V_0(\vec{r} - \vec{r}_n)$.

$V_0(\vec{r} - \vec{r}_n)$ is the potential energy of an electron when localized at the isolated atom positioned at r_n.

$V(\vec{r} - \vec{r}_n)$ represents the perturbation on H_0 due to the influence of atoms in the vicinity of $(\vec{r}_n)$, where the electron in question is strongly localized in relative terms and is written as

$$V(\vec{r} - \vec{r}_n) = \sum_{m \neq n} V_0(\vec{r} - \vec{r}_m)$$

Now, our aim is to look for the solution of the following Schrodinger equation:

$$H\psi_{\vec{k}}(\vec{r}) = \varepsilon_{\vec{k}}\psi_{\vec{k}}(\vec{r}) \tag{1.23}$$

where $\varepsilon_{\vec{k}}$ is the electron energy in the crystal, and $\psi_{\vec{k}}(\vec{r})$ is the Bloch wave function.

Let the solution of Equation (1.23) is of the following form:

$$\psi_{\vec{k}}(\vec{r}) = \sum_n \exp.\left(i\vec{k}.\vec{r}_n\right)\phi(\vec{r} - \vec{r}_n) \tag{1.24}$$

which satisfies the property of a Bloch function.

Now we calculate the energy of an electron with the wave vector $\vec{k}$ in the crystal based on the Equation (1.24)

$$\varepsilon_{\vec{k}} = E(\vec{k}) = \frac{\int \psi^*_{\vec{k}}(\vec{r})H\psi_{\vec{k}}(\vec{r})dV}{\int \psi^*_{\vec{k}}(\vec{r})\psi_{\vec{k}}(\vec{r})dV} \tag{1.25}$$

where H is the Hamiltonian operator for an electron in the crystal. The denominator takes care of the proper normalization of the Bloch function.

The denominator of Equation (1.25) with Equation (1.24) becomes

$$\int \psi^*_{\vec{k}}(\vec{r})\psi_{\vec{k}}(\vec{r})dV = \sum_{n,m} \exp.[i\vec{k}.(\vec{r}_n - \vec{r}_m)]\int \phi^*(\vec{r} - \vec{r}_m)\phi(\vec{r} - \vec{r}_n)dV \tag{1.26}$$

$\phi(\vec{r} - \vec{r}_m)$ has appreciable value only when the end point of the vector $\vec{r}$ lies in the vicinity of the atom $\vec{r}_m$. Therefore, we evaluate Equation (1.26) by putting $m = n$ in the first approximation, if there are N atoms in the crystal, we have

$$\int \psi^*_{\vec{k}}(\vec{r})\psi_{\vec{k}}(\vec{r})dV = \sum_n \int \phi^*(\vec{r} - \vec{r}_m)\phi(\vec{r} - \vec{r}_m)dV = \sum_n 1 = N \tag{1.27}$$

Where, we have assumed that the atomic wave functions were normalized.

Using Equations (1.21), (1.22), and (1.27), Equation (1.25) can be written as

$$\varepsilon_{\vec{k}} = E(\vec{k}) = \frac{1}{n}\sum_{n,m} \exp.\left[i\vec{k}(\vec{r}_n - \vec{r}_m)\right)\int \phi^*(\vec{r} - \vec{r}_m)[\varepsilon_0 + V(\vec{r} - \vec{r}_n)]\phi(\vec{r} - \vec{r}_n)dV \tag{1.28}$$

Here,

$$\int \phi^* \vec{r}_n V(\vec{r}-\vec{r}_n)\phi(\vec{r}-\vec{r}_n)dV = -\alpha$$

Where α is a constant which denotes an integral which can be evaluated. α is a small quantity, since the function $\phi(r - r_n)$ is appreciable only near the origin, where as $V(\vec{r}-\vec{r}_n)$ is small there.

Also,

$$\int \phi^*(\vec{r}-\vec{r}_m)V(\vec{r}-\vec{r}_n)\phi(\vec{r}-\vec{r}_n)dV = -\gamma$$

Where γ is the overlap integral, since it depends upon the overlap between orbitals centered at two neighbouring atoms. γ, though small, is non-vanishing because $V(\vec{r}-\vec{r}_n)$ is appreciable near the origin.

Thus, restricting the sum to nearest neighbours only, Equation (1.28) can be written as

$$E(\vec{k}) = -\alpha - \gamma \sum_{n,m} \exp \cdot \left[i\vec{k}(\vec{r}_n - \vec{r}_m) \right]$$

This gives the band energy as a function of $\vec{k}$.

The tight-binding approximation assumes that the interactions between neighboring atoms decreases quickly with increasing distance. Mathematically, this means the off-diagonal matrix elements of the Hamiltonian are approximated away as negligible, except for those corresponding to atoms that are close by. Thus, these matrix elements can be thought of as the probability of an electron hopping from one atom to another. The tight-binding approximation is easy to understand and results in a simple Hamiltonian thus giving fast calculations and helps in describing lots of materials.

1.5 LOW-DIMENSIONAL MATERIALS

Low-dimensional physics have revolutionized semiconductor physics, as the composition of the material can be changed on the scale of a nanometer. In true sense, the electrons in a material are free to move in three dimensions but can be made to move freely along fewer dimensions. Researchers have made this possible by trapping them in a narrow one, two- and three-dimensional potential well that restricts their motion in one-, two- and three-dimensions giving rise to two-dimensional, one-dimensional, and zero-dimensional materials respectively. The properties of low-dimensional systems are quite different from the bulk behavior. When the dimensions of the sample becomes comparable to the wavelength of the important excitation in a solid, a phenomenon known as "quantum size effect" occurs; for example, nanosized crystals exhibit behavior intermediate between that of bulk materials and molecules. Furthermore, if the size of the nanocrystal is comparable with the effective Bohr diameter of an electron in bulk crystal, then their band gap can be tailored according to their size. The diameter of the semiconductor crystal approaches the exciton Bohr diameter, its electronic properties start to change. For such materials, a blue shift in the optical band gap is observed. Furthermore, the physicochemical characteristics

TABLE 1.1
Typical Dimensions of 2D, 1D, and 0D Nanostructures

Nanostructure	Typical Nanoscale Dimensions
Thin film, quantum well (two-dimensional)	1–1,000 nm (thickness)
Quantum wire, nanowire, nanorods, and nanopillars (one-dimensional)	1–100 nm (radius)
Nanotubes	1–100 nm (radius)
Quantum dots, nanodots (zero-dimensional)	1–10 nm (radius)
Porous nanomaterials, aerogels	1–50 nm (particle size)
Sculptured thin films	10–500 nm

of a solid significantly change when one or more dimensions of a solid material are reduced to the scale of nanometer.

This small size gives rise to novel unique electrical and magnetic properties, and the reduced dimensionality material is called a low-dimensional structure (or system). For comparison, Table 1.1 gives the typical dimensions of the 2D, 1D, and 0D nanostructures. The low-dimensional materials are being classified on the basis of the number of reduced dimensions they have. In 3D structure or bulk materials, the particle is free to move in all directions, whereas for 2D material, the particle motion is restricted in one direction but is free to move in the other two directions; in 1D structure, the particle motion is restricted in two directions, leading to free movement along only one direction; and in 0D structure, the particle motion is restricted in all three directions.

In mechanics, quantum mechanics is used to solve the model systems that illustrate the confinement of a particle in a box, cylinder, and sphere, and the wave functions and energies of the confined particles can be obtained by solving Schrodinger's equation, and the difference between these systems lies in their boundary conditions. These constraints dictate where the particle can and cannot be, and reflect the underlying physical geometry and restrictions of the system. This is why one often refers to a box, a cylinder, and a sphere while modeling quantum well, quantum wires, and quantum dots, since there are natural geometries.

Nowadays, people are trying to design nanoelectronic devices based on quantum wells, quantum wires, and quantum dots. In these systems, the transfer of electrons is controlled. Therefore, for basic fundamentals with regard to nanoelectronics, it requires the understanding of physics for these three quantum structures.

1.6 QUANTUM CONFINEMENT IN LOW-DIMENSIONAL MATERIALS

The boiling point of water is determined on the basis of the average value of the behavior of billions and billions of molecules of water, and we assume that the result should be true for any size group of molecules. This is not true for nanomaterials, and there is a possibility that the same material can show different properties. This is because the same material follows the quantum physics rather than Newtonian physics. According to Newtonian physics, a body cannot be found on the other side of the wall if it does not have sufficient energy to pass it, and therefore the probability of finding the barrier is null. But in quantum mechanics, there is a finite probability of finding the particle on the other side of the barrier even if it does not have sufficient energy to jump the barrier, provided that the energy potential must be comparable to the wavelength of the particle. So, in simple words, it is possible for a particle with lower energy to exist on the other side of the energy barrier; of course, it is a violation of a fundamental law of classical mechanics. Furthermore, quantum mechanics provides a clear picture of the nature at the subatomic scale. It predicts all phenomena in terms of probabilities.

Nanomaterials are closer to the size of atoms or molecules; therefore, quantum mechanical laws are used to describe their behaviors. In nanomaterials, the electrons are confined in a small space rather than the small size of a bulk material, resulting in their confinement. It is referred to as quantum confinement. This is the main reason for many optical properties of nanomaterials, where quanta of light interact with the electrons of the nanomaterial. When the size of a material is reduced from macroscopic level, the properties remain the same at first and then small changes begin to occur. It should be noted that when the characteristics of a material change entirely from its bulk at a particular size in the range of nanoscale, the material is then called to be in the nanoscale range. Usually, it occurs at the size below 100 nm, and accordingly, scientists have developed terms to describe nanomaterials according to their shapes and sizes.

As stated earlier in quantum well structures (2D material), one dimension is reduced to the nanoscale range, whereas the other two dimensions remain large, and the material is simply a thin film of large (macroscopic) width and length but only few nanometers in thickness. Thus, the particles are said to be confined in one dimension (direction). Furthermore, in case of a quantum wire (1D material) two dimensions are in the nanoscale range while one remains large, and therefore can be thought as nanometer-sized cylinders that can measure several microns in length. Finally, for a quantum dot (0D material) all three dimensions are reduced to the nanometer range; therefore, confinement occurs in all three dimensions. All of these structures are shown in Figure 1.9.

The above statements are mentioned in Table 1.2.

The quantum confinement in 0D nanocrystals arises from the confinement of electrons within the crystallite boundary, which is analogous to the problem of particle in a quantum box. Furthermore, the properties of a material depend on the space available for the electrons for motion. The reduced size results in larger spacing in between the energy levels in the material. In the following subsections, we discuss the confinement of particle in a quantum well, quantum wire, and quantum dot separately.

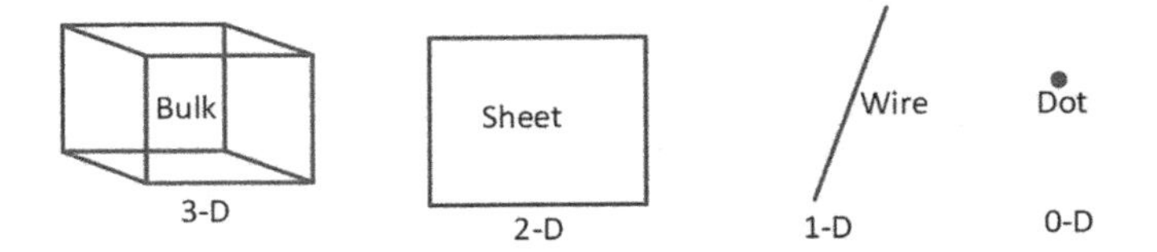

FIGURE 1.9 Three-, two-, one- and zero-dimensional sketches.

TABLE 1.2
Examples of Reduced Dimensionally Systems

Two-Dimensional Structure (1(z) Confinement)	One-Dimensional Structure (2(x, y) Confinement)	Zero-Dimensional Structure (3(x, y, z) Confinement)
Thin films	Nanorods	Nanoparticles
Quantum wells	Nanowires	Quantum dots
Grain boundary films	CNTs	Fullerenes

1.6.1 Particle Confinement in a Quantum Well

In a 2D material, the particles are confined to a thin sheet of thickness L_z along the z-axis by infinite potential barriers as shown in Figure 1.10. The structure so created is called a quantum well. As it is obvious from Figure 1.1, the particle cannot escape from the quantum well defined by $0 \le z \le L_z$.

Therefore, in the other two directions, i.e., x and y, the particle is free. Furthermore, it is assumed that the particle does not lose any energy on colliding with the walls of the well at $z = 0$ and $z = L_z$. It is the most powerful technique to control the properties of the materials. Such type of one-dimensional (1D) potential well is practically achieved by using two heterojunctions. Furthermore, it is the simplest confinement case in quantum mechanics.

After restricting the analysis to an infinitely deep 1D as shown in Figure 1.10, the potential will be of the following form:

$$V(z) = \begin{cases} 0, & 0 \le z \le L_z \\ \infty, & z \le 0 \text{ or } z \ge L_z \end{cases} \tag{1.29}$$

The time-dependent Schrodinger equation can be written as follows:

$$-\frac{\hbar^2}{2m}\frac{d^2}{dz^2}\psi(z) + V(z)\psi(z) = E\psi(z) \tag{1.30}$$

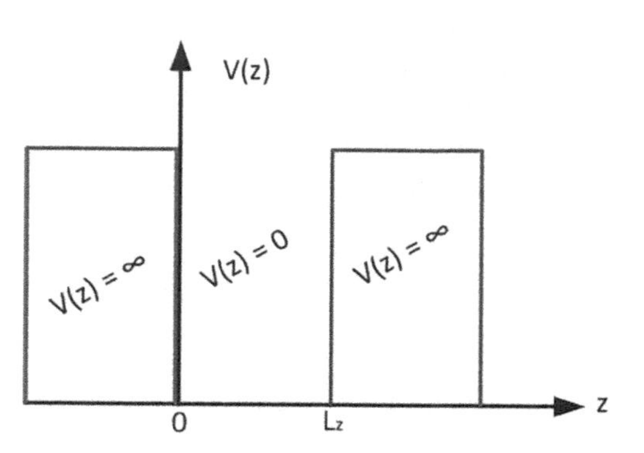

FIGURE 1.10 Schematic representation for the confinement of particle in a quantum well structure.

Outside the chosen potential well, i.e., $z \leq 0$ and $z \geq L_z$, the potential is infinite; hence, the only solution possible to Equation (1.30) is $\psi(z) = 0$, which simply means that all energy values are allowed outside the potential well.

With the help of Equation (1.29) within the potential well, Equation (1.30) reduces to

$$-\frac{\hbar^2}{2m}\frac{d^2}{dz^2}\psi(z) = E\psi(z) \quad 0 \leq z \leq L_z \tag{1.31}$$

It is to mention that the wave function $\psi(z)$ of the particle is zero on both walls and continuous inside the wall. Furthermore, the particle must exist somewhere in the well, and therefore, the probability of finding the particle inside the well must be one, i.e.,

$$\int_{-\infty}^{\infty} |\psi(z)|^2 \, dz = 1$$

with these stipulations, the solutions of the Equation (1.31) are many, and these solutions are called Eigen functions and may be written as

$$\psi_{n_z} = \sqrt{\frac{2}{L_z}} \sin \frac{n_z \pi}{L_z} z \qquad \begin{matrix} 0 \leq z \leq L_z \\ n_z = 1,2,3\ldots \end{matrix} \tag{1.32}$$

The complete solution to Equation (1.31) is the superposition of all the eigen functions and is given by

$$\psi(z) = \sum_{n=1}^{\infty} A_n \psi_{n_z}(z) \quad 0 < z < L_z \tag{1.33}$$

where A_n is the coefficient of expansion that determines the importance of each eigenfunction in the solution. Furthermore, each eigen function describes the state of electron confinement. The Eigen energy associated with n_zth eigenfunction is given by

$$E_{n_z} = \frac{\hbar^2}{2m}\left(\frac{n_z \pi}{L_z}\right)^2 \tag{1.34}$$

where $n_z = 1,2,3\ldots$

1.6.2 Particle Confinement in a Quantum Wire

In case of a one-dimensional (1D) structure usually called quantum wire (tubes, rods, belts, etc.), it is possible to decouple the motion along the length of the wire, which is taken to be along the x-axis as shown in Figure (1.11).

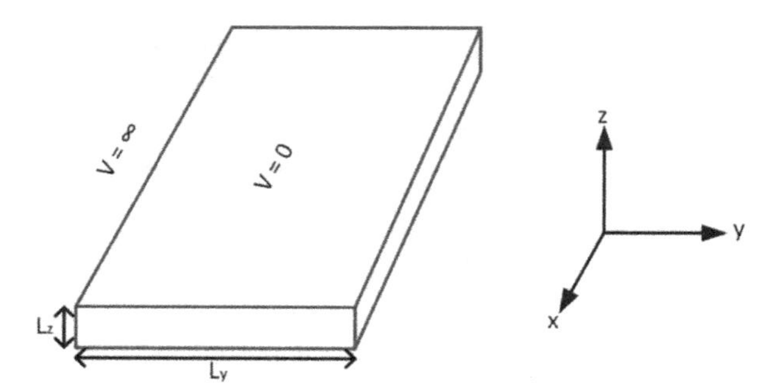

FIGURE 1.11 Schematic representation for the confinement of a particle in a one-dimensional (1D) structure.

The confinement of the particle is along the *y-z* plane. Thus, the potential *V(r)* is written as the sum of a two-dimensional confinement potential (*y-z* plane) and a potential along the length of the wire (*x*-axis):

$$V(\vec{r}) = V^{(1)}(x) + V^{(2,3)}(y,z) \tag{1.35}$$

Similarly, the wave function is written as the product of the two components:

$$\psi(\vec{r}) = \psi^{(1)}(x)\psi^{(2,3)}(y,z) \tag{1.36}$$

We have the time-dependent Schrodinger equation given as

$$\nabla^2\psi(\vec{r}) + V(\vec{r})\psi(\vec{r}) = E\psi(\vec{r}) \tag{1.37}$$

Thus, Equation (1.37) can be written as

$$\left\{-\frac{\hbar^2}{2m}\left(\frac{\partial^2}{\partial x^2}+\frac{\partial^2}{\partial y^2}+\frac{\partial^2}{\partial z^2}\right)+V^{(1)}(x)+V^{(2,3)}(y,z)\right\}\psi^{(1)}(x)\psi^{(2,3)}(y,z)$$

$$= E\psi^{(1)}(x)\psi^{(2,3)}(y,z) \tag{1.38}$$

The above equation can be split into the following two autonomous equations of motion:

$$-\frac{h^2}{2m}\frac{d^2}{dx^2}\psi^{(1)}(x) = E^{(1)}\psi^{(1)}(x) \tag{1.39}$$

$$-\frac{h^2}{2m}\left(\frac{\partial^2}{\partial y^2}+\frac{\partial^2}{\partial z^2}\right)\psi^{(2,3)}(y,z)+V^{(2,3)}(y,z)\psi^{(2,3)}(y,z) = E^{(2,3)}\psi^{(2,3)}(y,z) \tag{1.40}$$

The above equation is satisfied by a plane wave of the following form:

$$\psi^{(1)}(x) \sim e^{ik_x x} \tag{1.41}$$

where k_x is the wave vector along the x-axis, thus leading to the dispersion relation, given by

$$E^{(1)} = \frac{\hbar^2 k_x^{\ 2}}{2m} \tag{1.42}$$

This gives the energy levels along the x-axis in which direction the particle is free to move.

The potential $V^{(2,3)}(x,y)$ is given by

$$V^{(2,3)}(x,y) = \begin{cases} 0 & (0 < y < L_y) \cap (0 < z < L_z) \\ \infty, & \text{otherwise} \end{cases} \tag{1.43}$$

Outside the rectangular region, the wave function $\psi^{2,3}(x,y)$ is identically zero,

Thus, Equation (1.40) can be written as

$$-\frac{\hbar^2}{2m}\left(\frac{\partial^2}{\partial y^2} + \frac{\partial^2}{\partial z^2}\right)\psi^{(2,3)}(y,z) = E^{(2,3)}\psi^{(2,3)}(y,z) \tag{1.44}$$

The wave function $\psi^{(2,3)}(y,z)$ can be decomposed into two separate parts:

$$\psi^{(2,3)}(y,z) = \psi^{(2)}(y)\psi^{(3)}(z)$$

We can use the method of separation of the variables for energy as well; therefore, the above equation can also be written with respect to energy as follows:

$$E^{(2,3)} = E^{(2)} \,\&\, E^{(3)}$$

This leads to the following decoupled equations:

$$-\frac{\hbar^2}{2m}\left(\frac{\partial^2}{\partial y^2}\right)\psi^2(y) = E^{(2)}\psi^{(2)}(y) \tag{1.45}$$

$$-\frac{\hbar^2}{2m}\left(\frac{\partial^2}{\partial z^2}\right)\psi^2(z) = E^{(3)}\psi^{(3)}(z) \tag{1.46}$$

The above equations are identical to the Schrodinger equation in the deep potential well and therefore are subjected to similar boundary conditions. Actually, the potential energy outside the wire is infinite, and the boundary conditions of the continuity of the wave functions at the walls imply that the product $\psi_{(y)}^{(2)}$ and $\psi_{(z)}^{(3)}$ must be zero on the walls. Hence, eigensolutions are

$$\psi_{ny}^{(2)}(y) = \sqrt{\frac{2}{L_y}} \sin\left(\frac{n_y \pi y}{L_y}\right) n_y = 1,2,3,\ldots$$

$$\psi_{nz}^{(3)}(z) = \sqrt{\frac{2}{L_y}} \sin\left(\frac{n_z \pi z}{L_z}\right) n_z = 1,2,3,\ldots$$

$$\text{Thus, } \psi_{n_y,n_z}^{(2,3)}(y,z) = \sqrt{\frac{4}{L_y L_z}} \sin\left(\frac{n_y \pi_y}{L_y}\right) \sin\left(\frac{n_z \pi_z}{L_z}\right) \tag{1.47}$$

As in the case of a quantum well structure, the given energy for a quantum wire is inversely proportional to its size and mass. The corresponding energy levels are given by

$$E_{n_y} = \frac{\hbar^2}{2m}\left(\frac{n_y \pi}{L_y}\right)^2 \quad n_y = 1,2,3,\ldots$$

$$E_{n_z} = \frac{\hbar^2}{2m}\left(\frac{n_z \pi}{L_z}\right)^2 \quad n_z = 1,2,3,\ldots$$

where n_y and n_z are principle quantum numbers.

1.6.3 Particle Confinement in a Quantum Dot

A quantum dot is a quantum-confined structure in which the motion of the carriers is confined in the directions of *x*, *y*, and *z* by potential barriers. Such a structure is narrow in all three dimensions. The particle is not free at all. Generally, quantum dots are extremely small structures ranging from 2 to 10 nm in diameter. The radius of quantum dots is smaller than the characteristic de-Broglie wavelength or Bohr radius of carriers; that is, electrons/holes in the material and as such the carriers do not possess any degree of freedom. The quantum dot structure is usually designated as the quantum box as shown in Figure 1.12. It is a generalization of a quantum wire in which there is an additional confinement along the third axis.

There are other shapes of quantum dots, such as spherical quantum dots. As per Heisenberg's uncertainty principle, increased spatial confinement results in increased energy of the confined states.

The potential is of the form

$$V(x,y,z) = \begin{cases} 0 & \{0 < x < L_x\} \cap \{0 < y < L_y\} \cap \{0 < z < L_z\} \\ \infty & \text{otherwise} \end{cases} \tag{1.48}$$

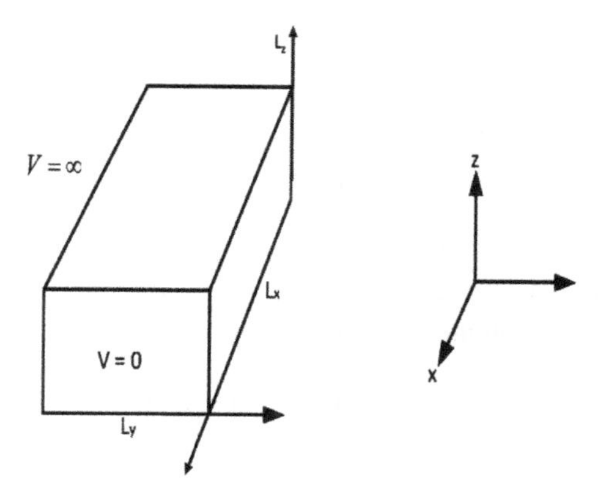

FIGURE 1.12 Schematic representation for the confinement of particle in a quantum dot (box) structure.

With this condition, the three-dimensional Schrodinger equation within the quantum box becomes

$$-\frac{\hbar^2}{2m}\left(\frac{\partial^2}{\partial x^2}+\frac{\partial^2}{\partial y^2}+\frac{\partial^2}{\partial z^2}\right)\psi(x,y,z)=E\psi(x,y,z) \tag{1.49}$$

The eigenfunction described by three principal quantum numbers n_x, n_y, and n_z is given by

$$\psi_{n_x,n_y,n_z}(x,y,z)=\sqrt{\frac{8}{L_xL_yL_z}}\sin\left(\frac{n_x\pi x}{L_x}\right)\sin\left(\frac{n_y\pi y}{L_y}\right)\sin\left(\frac{n_z\pi z}{L_z}\right) \tag{1.50}$$

where, $n_x = n_y = n_z = 1,2,3....$

Therefore, the Eigen energy for specifying Eigen function is given by

$$E_{n_x,n_y,n_z}=\frac{\hbar^2\pi^2}{2m}\left(\frac{n_x^2}{L_x^2}+\frac{n_x^2}{L_x^2}+\frac{n_x^2}{L_x^2}\right) \tag{1.51}$$

where, $n_x = n_y = n_z = 1,2,3....$

Here, in this case, the E_{n_x,n_y,n_z} is the total particle energy of a particle in a quantum dot structure. The discrete energy spectrum and the lack of free propagation are the main features of a quantum dot which distinguishes it from quantum wells and quantum wires. Furthermore, quantum dots are often called artificial atoms, as these features are typical for atoms.

1.7 DENSITY OF STATES IN BULK MATERIALS

The density of states is defined as the number of different states at a particular energy level that electrons are allowed to occupy. It is a measure of how closely the energy levels are to each other. It is defined as the number of available electron states per unit volume (dN) at energy "E" in the energy interval E and $E + dE$; then the DOS is given by

$$g(E) = \frac{dN(E)}{dE}$$

Furthermore, for determining the electronic properties like absorption emission and general distribution of electrons in a material, it is important to know the number of available states per unit volume per unit energy range. Moreover, the DOS depends upon the dimensionality of the system.

Now, let us assume a valley in the conduction band with energy level described by a parabolic relation with some effective mass m_c,

$$E(\vec{k}) = E_c + \frac{\hbar^2 k^2}{2m_c} \tag{1.52}$$

Let us consider a box of size $L_x L_y L_z$ with periodic boundary conditions in all three dimensions:

$$k_x = \frac{2\pi}{L_x} n_x,\; k_y = \frac{2\pi}{L_y} n_y,\; kz = \frac{2\pi}{L_z} n_z \tag{1.53}$$

where, n_x, n_y, n_z are integers.

The box is also considered to be so large that the allowed k-values are continuous, so that the summations over these indices can be replaced by integrals as

$$\sum_{k_x} \rightarrow \int_{-\infty}^{\infty} \frac{dk_x}{2\pi/L_x} \quad \sum_{k_y} \rightarrow \int_{-\infty}^{\infty} \frac{dky}{2\pi/L_y} \quad \sum_{k_z} \rightarrow \int_{-\infty}^{\infty} \frac{dk_z}{2\pi/L_z} \tag{1.54}$$

Therefore, the total number of states allowed $N(k)$ up to a maximum value of k is given by

$$\frac{L_x L_y L_z}{8\pi^3} \frac{4\pi k^3}{3} = \frac{k^3 \Omega}{6\pi^2} \tag{1.55}$$

Using Equation (1.52), $N(k)$ can be converted to $N(E)$ giving the total number of states which have energy less than E. The derivative of this function gives the DOS in the bulk material as

$$g(E) = \frac{1}{2\pi^2} \left(\frac{2m}{\hbar^2} \right)^{\frac{3}{2}} E^{\frac{1}{2}} \tag{1.56}$$

$$\text{i.e. } g(E) \propto E^{\frac{1}{2}}$$

The density of bulk material is a possible function of its energy. Therefore, $g(E)$ increases with increase in energy of the system.

1.8 DENSITY OF STATES IN 2D, 1D, AND 0D MATERIALS

1.8.1 Density of States in 2D Materials

In 2D materials, the electron motion is confined along one direction and free to move in other two directions. Therefore, along the confined direction (say *Z*), the energy is quantized, and along the other two directions (say *x*, *y*), the energy is not quantized and the electron can move like a free particle. Let us consider that the electrons are confined by infinite potential barriers at $Z = 0$ and $Z = d$ in a thin film of thickness "*d*" as shown in Figure 1.13.

The motion of the electrons in the *x*-*y* plane is assumed to be unconfined using a free electron approach. Here, we consider the effective mass "m^*" of the electron instead of quantum well structures free electron mass by taking into account the possible band structure. The motion of electron in *x*-*y* plane is described by wave vectors k_x and k_y along *X* and *Y* directions whose values are given by

$$k_x = \frac{2\pi}{L_x} n_x \text{ and } k_y = \frac{2\pi}{L_y} n_y$$

The total energy associated with the particle is the sum of energy along the quantized direction (k_z) and the energy along the other two free directions, i.e.,

$$E = \frac{\hbar^2}{2m^*} k_z^{\,2} + \frac{\hbar^2}{2m^*}(k_x^2 + k_y^2)$$

$$\text{Area of each mode } (k \text{ state}) = k_x \times k_y = \frac{(2\pi)^2}{L_x L_y} = \left(\frac{2\pi}{L}\right)^2$$

Let us consider the area of *k* space as

$$A_k = \pi k^2$$

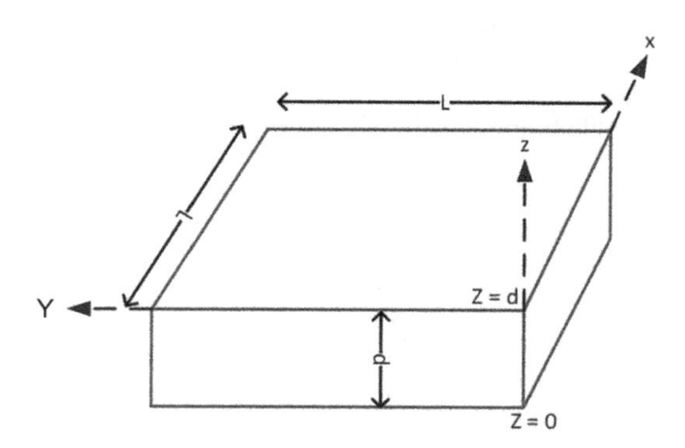

FIGURE 1.13 Schematic representation for a quantum well structure.

Therefore, the total number of modes in the area A_k is given by

$$T_m = \text{total area/area of each mode}$$

$$T_m = \frac{\pi k^2}{(2\pi^2)} \cdot L^2 = \frac{k^2 L^2}{4\pi}$$

If the particle be an electron, then there can be two electrons corresponding to the same energy, one with spin up and other with spin down. Therefore, the total number of energy states in the area πk^2 are given by

$$T_m = \frac{2k^2}{4\pi^2} L^2 = \frac{k^2 L^2}{2\pi^2}$$

Number of states per unit area is given by

$$N = \frac{\text{Total number of states}}{\text{Area}}$$

$$N = \frac{k^2 L^2}{2\pi L^2} = \frac{k^2}{2\pi}$$

Or

$$dN = \frac{k}{\pi} dk$$

Therefore, DOS is given by

$$g(E) = \frac{dN}{dE}$$

$$g(E) = \frac{k}{\pi} \frac{dk}{dE}$$

Along the free directions k_x and k_y, the energy is given by

$$E = \frac{\hbar^2 k^2}{2m}$$

$$k = \sqrt{\frac{2m}{\hbar^2}} E^{\frac{1}{2}}$$

$$dk = \sqrt{\frac{2m}{\hbar^2}} \left(\frac{1}{2} E^{\frac{-1}{2}} \right)$$

$$\therefore g(E) = \frac{k}{\pi}\sqrt{\frac{2m}{\hbar^2}}\left(\frac{1}{2}E^{\frac{-1}{2}}\right)$$

$$= \frac{1}{\pi}\left(\frac{2m}{\hbar^2}\right)^{\frac{1}{2}} E^{\frac{1}{2}}\left(\frac{2m}{\hbar^2}\right)^{\frac{1}{2}} \frac{1}{2}E^{\frac{-1}{2}}$$

$$g(E) = \frac{1}{\pi}\frac{2m}{\hbar^2}\frac{1}{2} = \frac{m}{\pi\hbar^2}$$

This is the energy density of the sub-bands for a given k_z (or E_n). For each successive k_Z, there will be an additional $\frac{m}{\pi\hbar^2}$ term, hence another sub-band. The $g(E)$ are therefore expressed as

$$g(E) = \frac{m}{\pi\hbar^2}\sum_n \Theta(E - E_n)$$

where $\Theta(E - E_n)$ is the Heaviside step function or unit step function and is given by

$$\Theta(E - E_n) = \begin{cases} 0 & \text{for } E < E_n \\ 1 & \text{for } E > E_n \end{cases}$$

Clearly in each sub-band, the DOS is a constant, i.e., independent of energy.

The DOS in a quantum well system is shown in Figure 1.14, where the DOS associated with all bound states has been added together to give a total DOS.

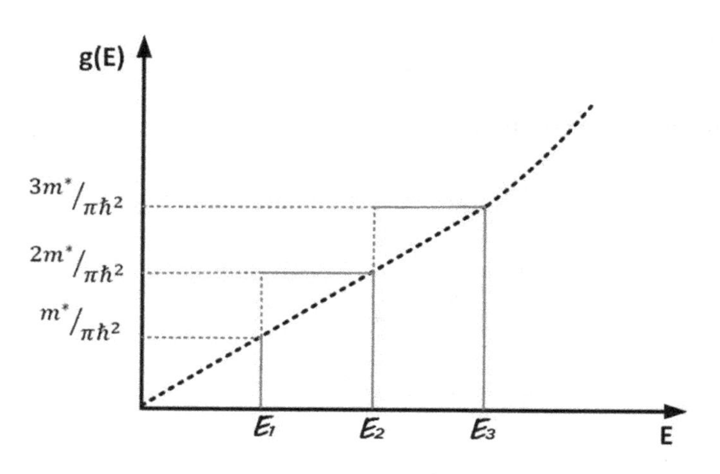

FIGURE 1.14 DOS in a quantum well.

1.8.2 Density of States in 1D Systems

In a one-dimensional (1D) material such as a quantum wire, the particle motion is confined along two directions and free along one direction. Let x, y directions are confined directions, and z is free direction for the particle. Therefore, the particle energy is quantized along x and y directions, and thus there is only one degree of freedom. The total energy of the particle is given by

$$E = E_x + E_y + E_z$$

$$E = \frac{\hbar k_x^{\ 2}}{2m} + \frac{\hbar k_y^{\ 2}}{2m} + \frac{\hbar k_z^{\ 2}}{2m}$$

$$\text{or } E = E_m + E_n + E_Z$$

where $k_z = \frac{2\pi}{L_z}$, and for the confined direction

$$k_x = \frac{m\pi}{L_x} \text{ and } k_y = \frac{n\pi}{L_y}$$

where m, $n = 1,2,3\ldots$ are integers.

We consider a length of $2k$ in K-space, such that the number of states along this length is given by

$$N_{2k} = \frac{2k}{k_z} = \frac{2k}{2\pi/L_z} = \frac{k}{\pi} L_z$$

If we consider an electron as a free particle, then there can be two electrons: one with spin up and other with spin down corresponding to each energy.

Therefore, the number of states along the length $2k$ is

$$N_{2k} = \frac{2k}{\pi} L_z$$

Therefore, the number of states per unit length is

$$N = \frac{N_{2k}}{L_z} = \frac{2k}{\pi}$$

$$E = \frac{\hbar^2 k^2}{2m}$$

$$k = \sqrt{\frac{2mE}{\hbar^2}}$$

$$\Rightarrow N = \frac{2}{\pi}\sqrt{\frac{2mE}{\hbar^2}}$$

Therefore, the DOS in 1D is given by

$$g(E) = \frac{dN}{dE}$$

$$g(E) = \frac{2}{\pi}\sqrt{\frac{2m}{\hbar^2}}\frac{d}{dE}(\sqrt{E})$$

$$g(E) = \frac{2}{\pi}\sqrt{\frac{2m}{\hbar^2}} \cdot \frac{1}{2}E^{\frac{-1}{2}}\frac{dE}{dE}$$

$$g(E) = \frac{1}{\pi}\sqrt{\frac{2m}{\hbar^2 E}}$$

This is the energy density for a given m, n values or E_m, E_n combination. From the above expression, the DOS in case of a one-dimensional material decreases with increase in energy for every (m, n) combination. Therefore, by taking into account all (m, n) combinations, the complete expression for DOS is given by

$$g(E) = \frac{1}{\pi}\sqrt{\frac{2m}{\hbar^2 E}}\sum \frac{1}{\sqrt{E_{K_z} - E_{m,n}}}\Theta(E_{K_z} - E_{m,n})$$

where $E_{m,n}$ are the confined energies associated with x and y directions, and $\Theta(E_{K_z} - E_{m,n})$ is the Heaviside or unit step function, such that

$$\Theta(E_{K_z} - E_{m,n}) = \begin{cases} 0 & \text{if } E_{k_z} < E_{m,n} \\ 1 & \text{if } E_{k_z} > E_{m,n} \end{cases}$$

It is to note that the peaks in the DOS at the sub-band threshold are called Van Hove singularities (Figure 1.15).

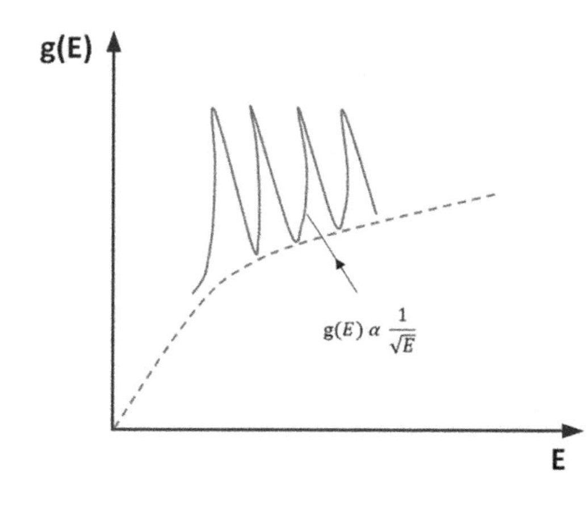

FIGURE 1.15 DOS in a one-dimensional structure.

1.8.3 Density of States in 0D Systems

In this case, the motion of a particle is confined along all the three directions (x, y, and z); that is, the particle is not free to move at all. Therefore, there is no dispersion curve, and the DOS depends on the number of confined levels. Therefore, one single isolated dot would offer two states (spin degenerate) at the energy of each confined level. The resultant material is called a quantum dot or nanoparticle. In this case, the energy is quantized along all three directions. Therefore, the total energy of the system is given by

$$E = \frac{\hbar^2 k_x^2}{2m} + \frac{\hbar^2 k_y^2}{2m} + \frac{\hbar^2 k_z^2}{2m}$$

$$\text{or } E = E_l + E_n + E_p$$

where, l, n, $p = 1,2,3\ldots$ are integers.

Also,

$$k_x = \frac{l\pi}{L_x}, k_y = \frac{n\pi}{L_y}, k_z = \frac{p\pi}{L_z}$$

It is clear that for a 0D system, there is no K-space to be filled with electrons and all available states exist only at discrete energies. Therefore, the DOS of 0D system is expressed by a delta function, i.e.,

$$g(E) = \delta(E - E_{l,n,p})$$

Figure 1.16 shows that the plots of $g(E)$ versus E for a zero-dimensional material.

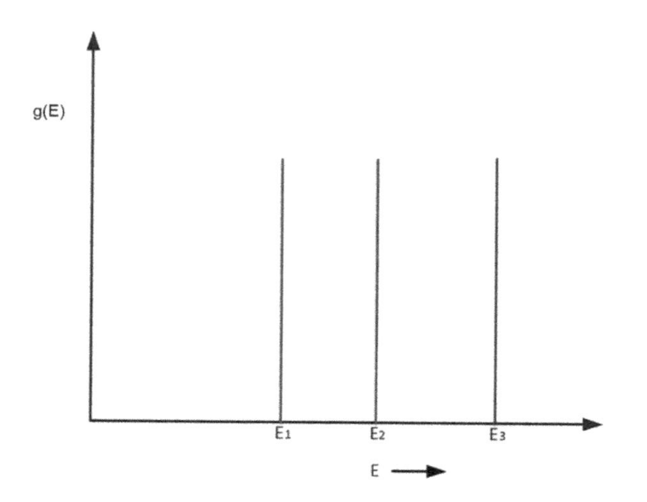

FIGURE 1.16 DOS of an ideal zero-dimensional (0D) system.

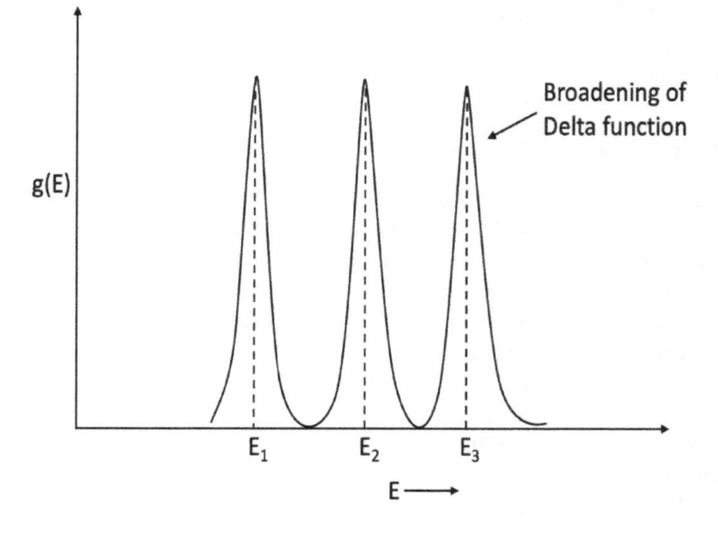

FIGURE 1.17 DOS of a real quantum dot.

In real quantum dots, the size distribution leads to a broadening of delta function as shown in ideal zero-dimensional system in Figure 1.17.

It is to mention that interaction between electrons and impurities as well as collisions with lattice vibrations brings a broadening of the discrete levels; however, for an idealized system, the peaks are very narrow and infinitely high.

1.9 EXAMPLES OF 0D, 1D, AND 2D MATERIALS

In nanostructure systems, there are three generic types of nanostructures: quantum dots (0D), quantum wires (1D), and quantum well (2D). These materials are referred to as low-dimensional systems as discussed in Section 1.5, where the dimensionality reflects the degree of freedom available for carriers in the material. Table 1.3 depicts the degree of confinement and degree of freedom for low-dimensional materials.

TABLE 1.3
Degree of Confinement and Degrees of Freedom for Low-Dimensional Materials

Structure	Degrees of Confinement	Degrees of Freedom
Quantum well	1	2
Quantum wire	2	1
Quantum wire	3	0
Bulk	0	3

Although there are hundreds of explored 0D, 1D, and 2D materials, we can group them under the banner of semiconductor system, metallic system, and carbon nanostructures.

1.9.1 Semiconductor Nanostructures

According to Moore's law the number of transistors for the same surface doubles for every eighteen months. This results in the replacement of Si devices with new fabricated materials such as thin epitaxial film and multi layered structures (2D) with versatile properties. These structures have led to the development of FETs, such as quantum well lasers, optical modulators, and quantum well photo detectors. Semiconductor nanostructures have attracted research over recent decades due to their unique physical and chemical properties and potential applications in the field of engineering, science, and technology. In general, the optical and electrical properties of semiconductor nanostructures can be tailored to a large extent by the dimensions of the crystal. Semiconductor nanostructures are adopted in a wide variety of electronic device applications including optoelectronic industry, solar cell technology, information technology. Among various semiconductor nanostructures, metal oxide nanostructures stand out as one of the most versatile materials due to their diverse properties and functionalities. These nanostructures exhibit exclusive properties which can be used in a variety of applications, but possess high anisotropic geometry and size confinement. The size-dependent properties of semiconductor nanostructures allows us to control the energy states available within a semiconductor device through dimension-induced band structure change, where electrons appear to have a quasi-continuum of available states.

Presently, the focus of development of semiconductor devices is on the synthesis techniques. In this regard, the development of different synthesis methods which may be capable of realizing high crystallinity and purity of materials is an enabling step towards making such nanoscale devices a reality. It is to mention that semiconductor nanowires have been made from other semiconductors such as Si; Ge; and most of the II–VI, III–V, and IV–VI semiconductors such as CdSe (cadmium selenide) nanowires and PbSe (lead selenide). The semiconducting nanoparticles have roughly the diameter of 5–10 nm. The semiconducting quantum dots and metallic quantum dots have the difference that the metallic nanoparticles have no distinct separation between the particles and the fringes depend upon particle orientation.

1.9.2 Metallic Nanostructures

These are equivalent to semiconducting nanostructures and the difference lies in the fact that most of the electrical properties of metallic systems can be described in a bulk-like fashion. Furthermore, the size of the metallic nanostructure should be somewhat smaller than semiconducting nanostructure, and the size-dependent change takes place in these materials; however, these are somewhat different in spirit than those of semiconducting nanostructures. A large number of metallic nanostructures have been fabricated during the past two decades using chemical- or template-based approaches with a mean diameter varying from 20 to 60 nm and lengths ranging from 2 to 20 μm, for example, silver nanowires, gold nanoparticles iron, and nanostructures. High-resolution TEM is often used to determine the crystal structure of the metallic nanostructure and reveals that the planes of atoms make up each particle.

1.9.3 Carbon Nanostructures

Here, the examples of carbon nanostructures include Bucky balls, carbon nanotubes (CNTs), graphene carbon nanofibers, and other 500 forms of carbon. C_{60} structurally resembles a soccer ball and was discovered by Harold Kroto, Richard Smalley, and Robert Curl in 1985 for which they were awarded the Nobel Prize in Chemistry in 1985. Since then, a number of variants such as C_{70} and C_{84} have been found. They are similar to zero-dimensional semiconducting quantum dots. These structures have a diameter of 1 nm. CNTs were discovered by S. Ijima in 1991 during arc evaporation synthesis. During his study, he found highly crystallized carbon filaments of few nanometers in diameter and few microns long called CNTs. They contain a chain of carbon atoms arranged in graphene sheets which are rolled together to form a seamless cylindrical tube. They are analogues to the nanowires and can be metallic or semiconducting depending upon their structure. There are two types of CNTs: (1) single-walled carbon nanotubes (SWCNTs) and (2) multi-walled carbon nanotubes (MWCNTs). The diameters of SWCNTs range from 0.4 to 2 nm, whereas those of MWCNTs range from 2 to 100 nm. There are three types of SWCNTs: (1) zigzag with $\theta = 0°$; (2) armchair with $\theta = 30°$; and (3) chiral with $0 < \theta < 30°$, where θ describes the direction of cross section of a tube relative to high symmetry direction in the sheet. The structure of the nanotube is described by (n, m) integers, which governs all their properties including electronic properties. However, it is to mention that to produce only one type of CNT is not practically possible till date. Furthermore, many efforts have been made in the past to separate CNTs according to their nature; however, there is little success.

Carbon nanofibers are a class of fullerenes that consist of curved graphene layers or nanocones stacked to form a quasi-one-dimensional filament, whose internal structure can be characterized by angle α between the graphene layers and the fiber axis. In the case of CNTs, $\alpha = 0$. Despite distinct difference in the internal structures between CNTs and nanofibers, nanofibers are often called nanotubes as they can display similar morphology to MWCNTs; however, their physical and chemical properties are different. Furthermore, the carbon nanofibers have two most commonly identified structural configurations: "herringbone type" in which dense conical graphene layers resemble a "fish skeleton" when viewed in cross section, and bamboo type in which cylindrical cup-like graphene layers alternate with cavities along the length like the cross section of the bamboo stem. While CNTs display ballistic electronic transport and diamond-like tensile strength along their axis, nanofibers have proven their robustness as individual free-standing structures with higher chemical reactivity and electronic transport across their side walls.

1.10 NON-EQUILIBRIUM GREEN'S FUNCTION (NEGF)

As the Schrodinger equation is known to be the basis of all quantum transport models, but solving it for a many particle system is very cumbersome process. Therefore, different methods like Green's function formalism is being used to appropriately analyze the nanosized devices. However, this technique only reformulates the problem giving no exact solution for a realistic device. Thus, the need for approximation techniques arises. The NEGF formalism proves to be an efficient conceptual and computational tool for analyzing quantum transport in nanodevices. It surpasses the Landauer method for ballistic, non-interacting electronics by including inelastic scattering and strong correlation effects at an atomistic level.

Device simulation software gives a self-consistent solution of transport equation and a Poisson equation iteratively till a converged value is reached as shown in Figure 1.18. In case of transport equation, electron density $n(r)$ and current I are obtained from a given potential $U(r)$, whereas the Poisson equation is used to calculate the effective potential $U(r)$ because of the other electrons present.

Consider a small device with one energy level ε with source and drain contacts. Applying a drain bias V will separate the common Fermi energy into two Fermi energies μ_1 and μ_2 as follows:

$$\mu_1 = E_f + \frac{qV}{2} \text{ and } \mu_2 = E_f - \frac{qV}{2} \tag{1.57}$$

For the device to be in equilibrium with source, the number of electrons would be equal to f_1, while as if the device would be in equilibrium with the drain, then the number of electrons would have been equal to f_2, such that

$$f_{1,2}(\varepsilon) = \frac{1}{\exp[(\varepsilon - \mu_{1,2})/K_B T]} \tag{1.58}$$

The actual number of electrons N lies in between the two Fermi energies given by the simple rate equations for the currents $I_{1,2}$ across the source and drain contacts:

$$I_1 = \frac{q\gamma_1}{\hbar}[f_1 - N] \text{ and } I_2 = \frac{q\gamma_2}{\hbar}[N - f_2] \tag{1.59}$$

where $\frac{q\gamma_1}{\hbar}$ and $\frac{q\gamma_2}{\hbar}$ show the rates at which an electron move from the source and the drain contacts, respectively.

Putting $I_1 = I_2 \equiv I$, the steady-state number of electrons N and the current I are given as

$$N = \frac{\gamma_1}{\gamma_1 + \gamma_2} f_1(\varepsilon) + \frac{\gamma_2}{\gamma_1 + \gamma_2} f_2(\varepsilon) \tag{1.60}$$

$$I = \frac{q}{\hbar} \frac{\gamma_1 \gamma_2}{\gamma_1 + \gamma_2} \left[f_1(\varepsilon) - f_2(\varepsilon) \right] \tag{1.61}$$

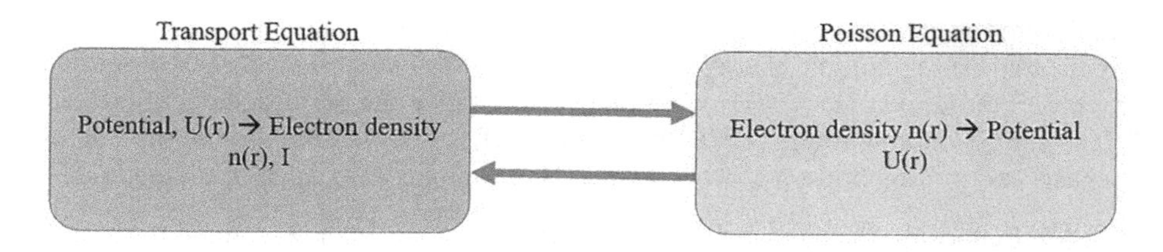

FIGURE 1.18 Self-consistent iteration of transport equation and Poisson-like equation.

This shows the current flow through a small conductor between the two contacts that try to maintain it between the two occupation levels f_1 and f_2. The real occupancy lies in between the two contacts (Equation 1.60). The source keeps pushing in the electrons and the drain emptying them, maintaining a continuous flow of electrons leading to a net current (Equation 1.61).

Broadening of discrete level into a distribution takes place due to the coupling between the source and the drain shown as

$$D(E) = \frac{\gamma/2\pi}{(E-\varepsilon-\Delta)^2+(\gamma/2)^2} \tag{1.62}$$

where γ is the line width with a possible shift in the level from ε to $\varepsilon+\Delta$.

Equations (1.60) and (1.61) can be rewritten for accommodating broadening terms as

$$N = \int_{-\infty}^{+\infty} dED(E)\left[\frac{\gamma_1}{\gamma_1+\gamma_2}f_1(E)+\frac{\gamma_2}{\gamma_1+\gamma_2}f_2(E)\right] \tag{1.63}$$

$$I = \frac{q}{\hbar}\int_{-\infty}^{+\infty} dED(E)\frac{\gamma_1\gamma_2}{\gamma_1+\gamma_2}\left[f_1(E)-f_2(E)\right] \tag{1.64}$$

Using the algebra, Equations (1.63) and (1.64) can be rewritten as

$$N = \int_{-\infty}^{+\infty} \frac{dE}{2\pi}\left[A_1(E)f_1(E)+A_2(E)f_2(E)\right] \tag{1.65}$$

$$I = \frac{q}{\hbar}\int_{-\infty}^{+\infty} \bar{T}(E)\left[f_1(E)-f_2(E)\right] \tag{1.66}$$

where

$$A_1 = G\gamma_1G^+, A_2 = G\gamma_2G^+, \bar{T} = \gamma_1G\gamma_2G^+ \tag{1.67}$$

$$G = \left[E-\varepsilon-\sigma_1-\sigma_2\right]^{-1}, \sigma_{1,2} \equiv \Delta_{1,2}-i\gamma_{1,2}/2 \tag{1.68}$$

In the above equations, it is assumed that the device has only one energy level "ε." But in the real sense, multiple energy levels exist. Thus, the device in general is described by the Hamiltonian matrix [H] whose eigenvalues give the permitted energy levels. This can be obtained by using different approaches like valence atomic orbitals as a basis and writing down a semi-empirical Hamiltonian with an ab initio method.

After choosing the basis, self-energy matrices $\left[\sum_{1,2}\right]$ can be defined for broadening and shift of energy levels. The scalar quantities like ε and $\sigma_{1,2}$ are replaced by the corresponding matrices [H] and $\left[\sum_{1,2}\right]$ in Equations (1.65) and (1.66), thus giving the appropriate NEGF equations. This gives

$$G=\left[EI-H-\sum\nolimits_{1}-\sum\nolimits_{2}\right]^{-1},\Gamma_{1,2}=i\left[\sum\nolimits_{1,2}-\sum\nolimits_{1,2}^{+}\right] \tag{1.69}$$

$$A_1(E)=G\Gamma_1G^+, A_2(E)=G\Gamma_2G^+ \tag{1.70}$$

where Γ is the identity matrix.

The number of electrons in Equation (1.65) is replaced by the density matrix given by the analogous quantity as

$$[\rho]=\int_{-\infty}^{+\infty}\frac{dE}{2\pi}\{[A_1(E)]f_1(E)+[A_2(E)]f_2(E)\} \tag{1.71}$$

The current is still given by Equation (1.66) after replacing the transmission as the trace of the analogous matrix quantity:

$$I=\frac{q}{\hbar}\int_{-\infty}^{+\infty}dE[f_1(E)-f_2(E)]Tr\left[\Gamma_1G\Gamma_2G^+\right] \tag{1.72}$$

which is exactly the Landauer formula for the current.

The equations shown above are the single-level scalar version (Equations 1.63–1.68) and the multilevel matrix version (Equations 1.66, 1.69–1.72) of NEGF equation. This is to give the introductive meaning of the quantities in NEGF equations. The NEGF equations also include the effect of incoherent scattering (not shown above) like electron–phonon interaction which becomes unavoidable as the channel length increases. This is done by defining additional self-energy matrices. Thus, NEGF provides a solid framework for analyzing new effects in transport models.

1.11 DENSITY FUNCTIONAL THEORY

DFT has been the most usable method for performing quantum mechanical simulations since the last three decades. It is one of the standard computational tools in condensed matter physics, chemistry, and biochemistry, and in both academic institutions and industrial establishments. The computational codes based on DFT are also used to investigate the structural, magnetic, and electronic properties of molecules, materials, and defects at different size scales.

In principle, the quantum mechanical wave function includes all the information about a desired system. In case of hydrogen atom or two-dimensional square potential, the Schrodinger equation can be solved exactly to obtain the wave function of the system. Then, the allowed energy states are determined. However, for many-body systems, it is impossible to solve the Schrodinger equation. Therefore, some approximations are needed to solve the problem and obtain the solution easily. This leads us to the simple definition of DFT, defining it as a method of finding an approximate solution to the Schrodinger equation in multibody systems. In a nut shell, DFT reduces 3N-dimensional problem to a three-dimensional one.

In case of many-body systems, the electrons get affected not only by the nuclei in their lattice sites but also by the other electrons present around them. Different types of interactions arise which are discussed as follows:

- The coulomb potential is a classical potential which arises due to the interaction between the systems of fixed electrons.
- The Hartree potential is due to the electron density distribution and ionic lattice, in which an electron moves independently feeling average electrostatic field due to all other electrons and the field due to the atoms.
- Exchange interaction arises due to the Pauli exclusion principle. An effective repulsion occurs between the electrons with parallel spins leading to spin interaction.
- Correlation interaction is due to the correlated motion of electrons with antiparallel spins arising because of the mutual coulombic repulsion.

Now in order to solve the many-body problem, different methods are used. In the Hartree–Fock method, the Hartree potential is used in addition to the exchange interaction which uses the antisymmetricity of the wave function. Parallel spins are kept at distance, thereby lowering the total binding energy of atoms. The disadvantage of this method is that it does not consider the correlation between the electrons with antiparallel spin. Many good approximations for ψ and E_0 are obtained readily, but such developments come at a very high computational cost and increase quickly with the number of electrons used. Also, exact solutions need a better description of the wave function, which also adds to the expenses of practical calculations. Thus, these limitations provide a good motivation for the use of DFT.

First, the number of degrees of freedom of the system should be reduced as much as possible. The Born–Oppenheimer approximation is used for this reduction of degrees of freedom. It also neglects the cross terms in the electronic wave function. It separates the wave function into two problems: first, the electron structure problem in which nuclei is considered as fixed in space and the solution is derived for the electronic degrees of freedom; and second, the nuclear problem in which solution is obtained for the nuclear degrees of freedom using potential energy surface (PES).

In 1964, Hohenberg and Kohn gave the two theorems known as the Hohenberg–Kohn theorem. The first theorem states that for any electronic system in external potential, ground state density determines such external potential uniquely. Thus, Hamiltonian is fully determined, which means many-body wave functions for all sates are determined. Therefore, if we only know the ground state density, all other properties of the system are completely determined or we can summarize it by saying the energy is a functional of the density. The second theorem gives a variational principle stating that the density, which is valid for external potential, can define the universal functional of energy [E], and for any particular external potential, the exact ground state of the system is determined by the global minimum value of this functional. It restricts the DFT to the studies of the ground state. Hohenberg–Kohn DFT is exact but impractical; therefore, in practical case, we use the Kohn–Sham approach.

The Kohn–Sham equations are similar to Hartree–Fock equations with the non-local exchange potential replaced by the local exchange-correlation potential. Kohn and Sham replaced the original particles with the fictitious system of non-interacting particles in an effective potential. The Kohn–Sham theory assumes the ground state density of interacting system is equal to that of some non-interacting system which is soluble with all terms included in some approximate functional of the density. The Kohn–Sham method gives an exact information of the density and ground state energy of a system. The computational cost of solving the Kohn–Sham equations is formally given as N^3 but in currently dropping towards N^1 through the exploitation of the locality of the orbitals. For the energy surface calculations, DFT is a practical and highly accurate alternative to the other wave function methods discussed above. Many approximations are used in the DFT based on the utility of the system in considerations. These approximations include

- Local density approximation (LDA)
- Generalized gradient approximation (GGA)
- Meta-GGA functionals
- Hybrid exchange functionals

The functionals currently utilized in the DFT form a natural hierarchy as shown in Table 1.4.

Developments are still going on in the underlying functional form, and thus the structure of ground state properties is advancing. The most notable development is the introduction of non-local nature of exchange potential in different forms.

DFT provides an efficient and unbiased way to calculate the ground state energy in different models of bulk materials and their surfaces. The quality and accuracy of

TABLE 1.4
Current Hierarchy of Exchange Correlation Functional

Dependencies	Family
Exact exchange	Hybrid
$\nabla^2\rho, \tau$	Meta-GGA
$\lvert\nabla\rho\rvert$	GGA
ρ	LDA

the acquired results depends on the development of approximations for the exchange-correlation energy functional. In recent years, many developments have been made in exchange-correlation functionals because of the local density gradient dependency, semi-local measures of the density, and introduction of non-local exchange functionals.

1.11.1 Advantages of DFT

- It is used in the calculation of total energies and forces for structure prediction, thermodynamics, and kinetics.
- It provides description of the electronic structure of the system, thus leading to observation of different properties like optical properties.
- It maintains healthy balance between accuracy or prediction and cost of computing, thus enabling calculations with thousands of electrons.
- It is a very well-established approach with all the capabilities and shortcomings understood.
- Highly efficient implementations are available for different types of problems.

1.12 SUMMARY

Low dimensional materials possess unique intrinsic properties, which notably depart from those of the bulk solid. The confinement of electrons or holes leads to a dramatic change in their behaviour and to the manifestation of size effects that usually under he category of quantum size effects. Furthermore, suitable control of the properties and responses of nanostructures can lead to new devices and technologies. The most significant nanostructure required to design nanoelectronic devices are quantum well, quantum wires and quantum dots. They are the basic building blocks of nanoscale electronic devices. All basic fundamentals of nanoelectronics require the understanding of physics of these three quantum structures.

G. Moore suggested that the number of transistors for the same surface doubles for every eighteen months, which is called "Moore's law". This results in the replacement of Si devices with new fabricated materials, such as thin epitaxial film and multilayered structures with versatile properties. These structures have led to the development of various nanoscale electronic devices. The properties of carbon nanotubes make them potentially useful in nanometer scale electronic device applications. They show unusual strength, unique electrical properties, and extremely high-thermal conductivity. Semiconductor nanostructures have attracted research over recent decades due to their unique physical and chemical properties, and potential applications in the field of engineering, science, and technology. In general, the optical and electrical properties of semiconductor nanostructures can be tailored to a larger extent by the dimensions of the crystal. Furthermore, semiconductor nanostructures are adopted in a wide variety of electronic device applications. Among various semiconductor nanostructures, metal oxide nanostructures stand out as one of the most versatile materials due to their diverse properties and functionalities. The size-dependent properties of semiconductor nanostructures allow the scientists and engineers to control their properties. In low-dimensional devices, the carriers are confined within the crystal, i.e., quantum wire, or in a low-dimensional potential well, i.e., quantum well devices. To enhance the functionality of semiconductor nanostructures/materials and reduced energy consumption, the major focus of development of semiconductor devices is on the synthesis techniques. The fabrication method is of significant importance because it determines the particle composition, structure, and size distribution.

REFERENCES

1. K. Kulkarni Sulabha, *Nanotechnology: Principles and Practices*, Capital Publishing 286 (New Delhi, 2007).
2. S.M. Lindsay, *Introduction to Nanoscience*, Oxford University Press, 480 (Oxford, 2009).
3. R. Kelsall, I. Hamley, M. Geoghegan, *Nanoscale Science and Technology*, John Wiley & Sons, 472 (Hoboken, NJ, 2005).
4. G.L. Hornyak, H.F. Tibbals, J. Dutta, J.J. Moore, *Introduction to Nanoscience and Nanotechnology*, CRC Press, 1640 (Boca Raton, FL, 2008).
5. J.H. Davies, *The Physics of Low Dimensional Semiconductors: An Introduction*, Cambridge University Press, 460 (Cambridge, 1998).
6. K.A. Shah, M.A. Shah, *Nanotechnology: The Science of Small*, Wiley, 200 (Hoboken, NJ, 2019).
7. W.A Goddard III, D. Brenner, S.E. Lyshevski, *Handbook of Nanoscience, Engineering, and Technology*, CRC Press, 1093 (Boca Raton, FL, 2012).
8. P.Y. Yu, M. Cardona, *Fundamentals of Semiconductors: Physics & Materials Properties*, Springer-Verlag, 617 (Berlin, 1996).
9. S., Datta, *Electronic Transport in Mesoscopic Systems*, Cambridge University Press, 377 (Cambridge, 1997).
10. N. Peyghambrain., S.W. Koch, A. Mysyrowicz, *Introduction to Semiconductor Optics*, Prentice Hall, 485 (Upper Saddle River, NJ, 1993).
11. M. Lundstrom, J. Guo, *Nanoscale Transistors, Device Physics, Modeling and Simulation*, Springer, 218 (Berlin, 2006).

2 Carbon Nanotubes and Their Device Applications

2.1 INTRODUCTION

Carbon nanotubes (CNTs) are sheets of graphite cells joined to look like a fence of lattice, which are then rolled up to form a tube-like shape. CNTs are of two forms: single-walled CNTs (SWCNTs) and multiple-walled CNTs (MWCNTs). The physical properties of SWCNTs differ significantly from those of MWCNT. Which type of CNT be chosen depends on the types of applications. SWCNTs have diameters ranging from 0.4 to 3 nm and MWCNTs from 2 to 100 nm, whereas double-walled carbon nanotubes (DWCNTs) have diameters close to 2 nm. There are three crystallographic configurations of ideal SWCNTs, zigzag, armchair, and chiral, which depend on how the graphene sheet is rolled up. In the zigzag conformation, two opposite C–C bonds of each hexagon are parallel to the tube axis, whereas in the armchair conformation, the C–C bonds are perpendicular to the axis, and in chiral configuration, the opposite C–C bonds lie at an angle to the tube axis. SWCNTs may be metallic or semiconducting in nature. The bandgap of SWCNTs is in the range from 0.4 to about 2 eV, whereas that of MWCNTs is zero [1,2].

The structure of SWCNT can be conceptualized by wrapping an atomic thick layer of graphite called graphene into a perfect cylinder. The way the graphite sheet is wrapped is represented by a pair of indices (n,m), called the chiral vector. The indices "m" and "n" denote the number of unit vectors along two directions in the honeycomb crystal lattice of graphene. The chiral vector of a CNT, which links two equivalent crystallographic sites, is given by $C_h = na_1 + ma_2$, where "a_1" and "a_2" are unit vectors of the graphene lattice, and the numbers "n" and "m" are integers. The set of integer numbers describes entirely the metallic or the semiconducting character of any CNT. If $m = 0$, the nanotubes are called zigzag. If $n = m$, the nanotubes are called armchair. Otherwise, they are called chiral. It is clear from Figure 2.1 that the CNT formed by the vectors is represented by (3, 1).

It is possible to recognize zigzag, chiral, and armchair CNTs just by following the pattern across the diameter of the tubes and analyzing their cross-sectional structure. In zigzag tubes, hexagons of carbon are arranged around the circumference so that an apex of the hexagon is parallel to the longitudinal axis of the SWCNT. Armchair SWNTs, however, have the apex oriented along the circumference of the tube and a C–C bond oriented along the longitudinal axis of the tube.

The difference between these types of CNTs depends on how the graphite is rolled up during its growth process. Each structure of CNTs in these three types is different, and each of them performs different roles. Each structure of CNTs has its own set of properties that make it appropriate for different uses across science,

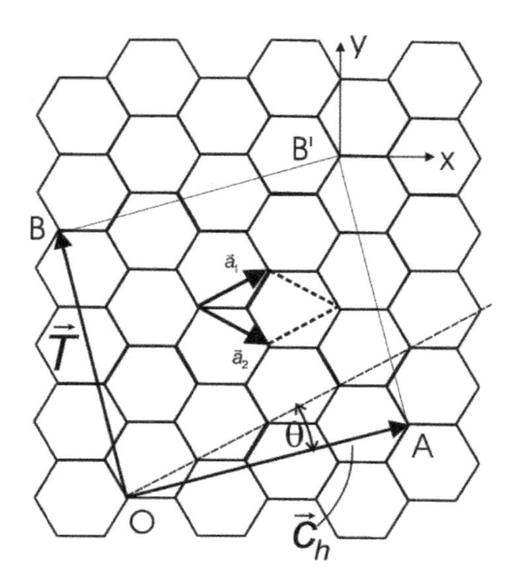

FIGURE 2.1 Construction of CNT from a graphene sheet.

architecture, geology, agriculture, and engineering. The physics of CNTs is extremely studied, and the mathematical description is quite well developed. The structure of CNTs is yet to be explored more, which will help in utilizing them in everyday applications. CNTs, especially SWCNTs, are one of the most remarkable discoveries of nanomaterials. Their physics and chemistry are both well understood. Their potential is enormous although it is worth to see how many of their commercially viable applications will be developed over the next decade.

CNTs may exhibit extraordinary aspect ratios and can be produced up to several centimeters long [3,4]. The CNT properties are strongly dependent on their structure, e.g., for typical diameters, all armchair SWCNTs and one-third of all zigzag nanotubes are metallic, and the rest are semiconducting [5]. Furthermore, both SWCNTs and MWCNTs possess a high surface area per unit weight, good mechanical properties, and high electrical conductivity at metallic state and high thermal conductivity/stability. It is to be kept in mind that for abovementioned applications, highly reliable synthesis techniques are required that should be capable of obtaining large quantities of high purity materials. Furthermore, controlled growth at precise lithographically patterned areas is required for applications in nanoelectronics and photonics for which understanding how to control the synthesis of CNTs is vital in order to deterministically integrate such nanostructures into various technologies. It is found that among various techniques developed for the synthesis of CNTs, catalyst chemical vapor deposition (CCVD) is one of the best to satisfy the abovementioned requirements. Although there are other techniques that are being used for the synthesis of CNT such as arc discharge and laser ablation, CVD is the most versatile and promising technique in terms of both bulk production and direct device integration.

2.2 PHYSICAL PROPERTIES OF CARBON NANOTUBES

CNTs are long, thin cylinders of carbon and different than other materials in size, shape, and properties. These structures have shown promising characteristics, and a lot of research has been made to understand various aspects of CNTs. Some of the unique equilibrium physical constants of single-walled CNTs are listed in Table 2.1.

The extraordinary properties of CNTs are to a large extent derived from their 1D character and the peculiar electronic structure of graphite. They have extremely low electrical resistance. CNTs have high stiffness and axial strength because of the carbon–carbon sp^2 bonding. Nanotubes are known to be the stiffest fibers, with a measured Young's modulus of 1.4 TPa. CNTs have been shown to have a thermal conductivity of at least twice that of diamond which was known to have the highest thermal conductivity. The specific heat and thermal conductivity of CNT systems are determined primarily by phonons. The measurements yield linear specific heat and thermal conductivity above 1 K and below room temperature. The linear temperature dependence of CNTs can be explained with the linear k-vector dependence of the frequency of the longitudinal and twist acoustic phonons. The measurements of the thermoelectric power of nanotube systems give direct information for the type of carriers and conductivity mechanisms.

In addition, CNTs carry as high as 10^9 A/cm^2. One application of nanotubes that has already been developed is extremely fine electron guns, which could be used as miniature cathode ray tubes (CRTs) in thin high-brightness low-energy low-weight displays. This type of display would consist of a group of many tiny CRTs, each providing the electrons to hit the phosphor of one pixel, instead of having one giant CRT whose electrons are aimed using electric and magnetic fields. These displays are known as field emission displays (FEDs). A nanotube formed by joining nanotubes of two different diameters end to end can act as a diode, suggesting the possibility of constructing electronic computer circuits entirely out of nanotubes. Nanotubes

TABLE 2.1
Equilibrium Physical Quantities of SWCNTs

S. No	Physical Quantity	Value
01	Average diameter of SWNTs	1.2–1.4 nm
02	Distance from opposite carbon atoms	2.83 Å
03	Analogous carbon atom separation	2.456 Å
04	Parallel carbon bond separation	2.45 Å
05	Carbon bond length	1.42 Å
06	C–C tight bonding overlap energy	~2.5 eV
07	Lattice constant	17 Å
08	Interlayer spacing of (n, m) chiral nanotube	3.39 Å
09	Density of (10,10) armchair nanotube	1.33 g/cm^3
10	Density of (12,0) zigzag nanotube	1.34 g/cm^3
11	Density of (12,6) chiral nanotube	1.40 g/cm^3

have been shown to be superconducting at low temperatures. CNTs are tubular carbon molecules provided with very particular properties. Their structure is similar to fullerenes. While fullerenes form a spherical shape, nanotubes are cylindrical structures with the ends covered by half of a fullerene.

Nanotube diameter is of the order of few nanometers, whereas its length is of the order of several millimeters. The physical properties make them potentially useful in nanometer-scale electronic and mechanical applications. Nanotubes show unusual strength, unique electrical properties, and extremely high thermal conductivity. The chemical bonding between carbon atoms inside nanotubes is always of sp^2 type. Nanotubes align themselves into ropes held together by the van der Waals forces and can merge under high pressure. They can be excellent conductors as well as semiconductors, depending on their structure. The thermal conductivity of CNTs is also high in the axial direction.

It is estimated that CNTs can carry a billion amperes of current per square centimeter. Copper wires fail at one million amperes per square centimeter because resistive heating melts the wire. One reason for the high conductivity of the CNTs is that they have very few defects to scatter electrons and thus a very low resistance. High currents do not heat the tubes in the same way that they heat copper wires. A significant result about semiconductor nanotubes shows that their energy gap depends upon the reciprocal nanotube diameter and is independent of the chiral angle of the semiconducting nanotubes given by

$$E_g = \{2a_{c\text{-}c}\ t\}/d_t \tag{2.1}$$

where $a_{c\text{-}c} = 1.44\,\text{Å}$ is the nearest C–C distance on a graphene sheet, $t = 2.5\,\text{eV}$ is the nearest neighbor C–C tight binding overlap energy, and $d_t = (n^2 + m^2 + nm)_{1/2}$ $0.0783\,\text{nm}$ is the tube diameter. It is clear from Equation (2.1) that the energy gap decreases as the diameter of a semiconducting tube increases. The electronic properties of one-dimensional (1D) conductors are of much interest because of their very rich phase diagram and the prediction that in a 1D system, the Coulomb interaction can lead to a strongly correlated electron gas called a Luttinger liquid instead of the weakly interacting quasi-particles described as a Fermi liquid in conventional metals. This issue is yet to be resolved. There are experimental results both for SWNTs and MWNTs, which speak in favor of either exotic Luttinger liquid or conventional Fermi liquid behaviors.

The mechanical properties of CNTs are also of large interest to scientists. Nanotubes may be called the ultimate fibers because of their light weight and the predicted record—high strength. They can be hundred times stronger than steel, while they are one-sixth its weight. There are no experiments done yet to confirm such a high-tensile strength. However, experiments have already demonstrated that nanotubes have an extremely high Young's modulus, on the order of TPa, and that they resist deformation remarkably well. When they are strongly bent, nanotubes do not break but "buckle," which means that the cylindrical structure locally flattens, just as what happens to a drinking straw. When the bending strain is released, the nanotube resumes its original straight shape. Nanotubes therefore appear to be ideal tips for scanning probe microscopes since they are not only very small but also

TABLE 2.2
Young's Modulus, Tensile Strength, and Density of CNTs Compared with Other Materials

Material	Young's Modulus (GPa)	Tensile Strength (GPa)	Density (g/cm^3)
Single-walled carbon nanotube	1,054	150	–
Multi-walled carbon nanotube	1,200	150	2.6
Steel	208	0.4	7.8
Epoxy	3.5	0.005	1.25
Wood	16	0.008	0.6

survive crashes with the sample surface. Table 2.2 gives the comparison of mechanical properties of CNTs with steel, epoxy, and wood.

CNTs have attractive mechanical and electronic properties. Nanotubes have a high mechanical strength due to a very large Young's modulus. To mention only a few ideas, nanotubes could be used as nanoscale electronic devices, as a balance to weigh small particles or as ultra-sharp and crash proof tips in a scanning probe microscope for the storage of hydrogen to store energy in electrochemical double-layer capacitors or to reinforce composite materials. A single nanotube can be used as a sensor, a nanorelay, a vessel, or a template. It is possible to produce light bulbs and fibers with CNTs. An array of CNTs can act as a flat panel display using their feature to act as field-emitting devices.

CNTs are considered as metallic or semiconducting in nature, and it is being said in a sample, one-third are metallic and two-third semiconducting. Nowadays, it is not possible to select the desired characteristic of a nanotube in advance. It is only possible to separate metallic from semiconducting tubes by using an electrical field. Metallic nanotubes with their diameter of a few nanometers represent the ultimate conducting wire, whereas the semiconducting ones can be used as transistors even on a transparent and flexible substrate. The transistors can be optimized by the chemical control of the nanotube–electrode interface. Quantum dots and spin valves can be built similar to simple logic gates and a Y-junction rectifier.

To analyze, we are thus getting used to the fact that nanosized carbon is not equal to ordinary carbon. We might still want to think though that gold is gold, platinum is platinum, and cadmium oxide is cadmium oxide, but we slowly have to get used to the fact that this is also not true, and the size and shape of the materials has a crucial role in its properties.

2.3 BALLISTIC TRANSPORT AND QUANTUM CONDUCTANCE IN CNTs

Under the ballistic transport phenomenon transport of electrons in a medium experience negligible or no scattering due to atoms or impurities [6], and when the length of the wire is less than the electronic mean free path, the electron transport in the wire is ballistic [7]. The conductance of the in the wire can be quantized, and each conduction channel has a conductance of

$$G_0 = \frac{2e^2}{h} \tag{2.2}$$

where G_0 is called "conductance quantum" and has the value of 112.9 $k\Omega$, e is the electron charge, and h is the Planck's constant.

Let us consider a one-dimensional wire which is connected adiabatically to two electrochemical potential reservoirs μ_1 and μ_2, and there are no reflections of electrons between the reservoirs. Furthermore, we assume that the wire is very narrow in order for the lowest transverse modes in the wire to be below the Fermi energy. As we know that the Fermi energy is the energy of the highest occupied quantum state in a system of fermions at absolute zero temperature.

The current is the same as current density in one dimension given as

$$J = ev\left(\mu_1 - \mu_2\right)\frac{dn}{d\varepsilon} \tag{2.3}$$

where $\frac{dn}{d\varepsilon}$ is the density of states, and v is the velocity of the electron.

With the inclusion of spin degeneracy, density of states is given as

$$\frac{dn}{d\varepsilon} = \frac{2}{hv} \tag{2.4}$$

The chemical potential difference between the reservoirs is an electron charge multiplies the voltage V across the reservoirs:

$$(\mu_1 - \mu_2) = -eV \tag{2.5}$$

Now, the total conductance using Equations (2.3), (2.4), and (2.5) is given as

$$G = \frac{J}{V} = \frac{2e^2}{h} \tag{2.6}$$

As per the Landauer formula, conductance is generally given as

$$G = \frac{2e^2}{h} T\ M(\mu) \tag{2.7}$$

where T is the transmission probability for a conducting channel, and $M(\mu)$ is the number of conduction channels and is a function of energy (μ). In this case, a conduction channel is characterized by the TM (μ) and contributes to the total conductance G by G_0TM (μ) [8]. For an ideal case (100% of transmission), a "wire" with one conduction channel (M = 1) has conductance of $G_0 = \frac{2e^2}{h}$ which is Equation (2.6).

Furthermore, G_0 is independent of properties and dimensions of a conductor. Hence, when the length of a wire is smaller than its electronic mean free path, its conductance changes only in unit of G_0 as conduction channels vary. Furthermore, as we know that the diameter of CNTs is very small, so the ratio between the diameter and the length of CNTs can reach as high as 1:132,000,000.

2.4 CNT TWO-PROBE DEVICES

SWCNTs are considered as potential building blocks for future nano-electronic devices. Different types of prototypical devices such as transistors, diodes, sensors, optoelectronic devices and non-volatile memory devices have been demonstrated using the SWCNT. The energy gap in semiconducting CNTs is inversely dependent on the nanotube diameter [9]. The weak scattering cross section and negligible electron–phonon interactions result in ballistic conduction of charge carriers in CNTs. It has been shown using femtosecond time-resolved photoemission calculations that the mean free path of metallic CNTs lies in the range of few micro meters [10]. Such a large mean free path is much above the normal lengths of CNT devices, in which the length of conducting channel fall in the submicron regime. Thus, CNTs can be considered for use in low power consumption electronic devices.

In nanostructure-based two-probe devices, different types of CNTs, molecules, and DNAs are used as channel for observing transport properties. All of these devices are one dimensional in nature having two ends. CNT two-probe devices have attracted lots of attention due to their potential applications in the electronic industry. CNT two-probe devices are like conventional two-terminal devices containing CNT as the transmission channel. These devices offer good electrical conductivity, thermal conductivity, and high chemical, thermal, and mechanical stability, which make them one of the most promising subjects of nanoelectronics.

CNT two-probe devices as the name suggests consist of a left electrode, a right electrode, and the central scattering region as shown in Figure 2.2. The two electrodes act as the input and output terminals called as probes. The two probes may be of same or different material as used in central scattering region. These devices have explained themselves as perfect candidates in nano-transistors, memristors, solar cells, and sensing devices.

Lots of work is being done to explore the basic architectures defining these two-probe-based devices. The two-probe structure is simple to construct and the

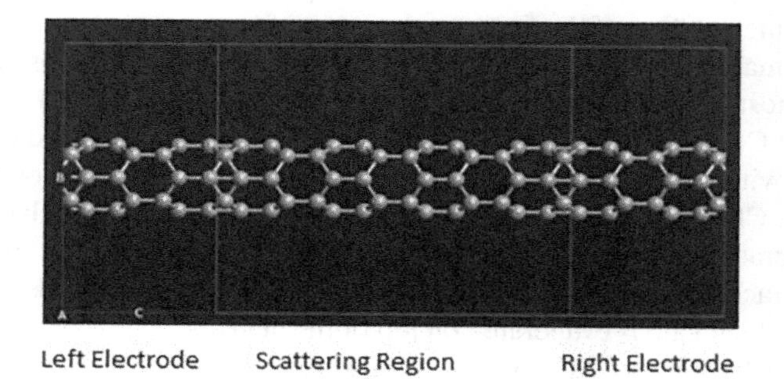

FIGURE 2.2 Model of pure two-probe CNT system consisting of two on tube electrodes and zigzag (4, 0) CNT as scattering region.

one-dimensional nature of CNTs makes their integration feasible in this device. It is worth to note that we should not get confused about the CNT two-probe devices and the CNT probes. CNT probes are simply CNTs attached to various devices and then used as tips in various imaging and sensing applications like atomic force microscopy (AFM).

The main contribution to the conductivity in these two-probe devices is made by electrons. As such, the conductivity of CNTs is determined by the mechanism of electron mobility. In a defect-less SWCNT at low temperatures, electrons pass through the nanotube without scattering, which corresponds to the quantum ballistic mechanism of conductivity. Electrical transport in CNTs has been intensively investigated experimentally and theoretically. Delocalized π-electron system is responsible for conduction in CNTs. SWCNTs are ballistic conductors at room temperature with two and one spin-degenerate conducting channel in metallic and semiconducting types, respectively. The channels belong to the first π- and π*-band of the delocalized π-electron system. Thus, at ambient conditions, without external excitation, SWCNTs only have two spin-degenerate conducting channels available for charge transport.

2.5 CNT FIELD-EFFECT TRANSISTORS (CNTFETs)

According to Moore's law, the number of devices in a circuit doubles every two years. This scaling down of devices has been the motivation among the engineers and scientists in the late 20th century. However, it has been seen that serious limitations arise as the critical dimension shrunk down to sub 22 nm range. The limits involve electron tunneling through short channels, associated leakage currents, passive power dissipation, short-channel effects and variations in device structure and doping. These limitations can be removed using a single CNT or an array of CNTs instead of bulk MOSFET structure. CNTs are being continuously studied for fast computing and low power consumption nanodevices. As it is expected, CNT-based FETs can overcome the scaling limitations of silicon FETs.

A carbon nanotube field-effect transistor (CNTFET) is a type of FET that utilizes a single CNT or number of CNTs as the channel medium. With the use of a semiconducting SWCNT, a FET has been built with prominent effects due to modulation of gate parameters. One of the first electronic nanodevices manufactured using CNT was the normal FET with a backgate. The design used at that time is still the basic step for almost all CNT-based nanodevices. The common CNTFET device design consists of a CNT lying on the top of two electrodes which are placed on the dielectric surface with a back gate. In newer designs of CNTFET, the electrodes are placed on top of the CNTs to improve contact stability with the same type of back gate used. The developments are going on to increase gate efficiency by using different techniques of attaching gates with the CNT channel. One of the methods include using topgate, in which gate is evaporated on top of the device.

At first, a back-gated FET-type structure was used with the CNT placed on a silicon substrate with oxidized surface. Source and drain electrodes were kept on the CNT structure itself. With the drain voltage fixed, the drain current has a strong dependence on the applied gate voltage V_G. This effect is caused due to the gate modulation which causes signal gain. Different fundamental properties of CNTFETs have been investigated like sub-band formation in the SWCNT channels [11], a long-channel CNTFET mechanism [12], transport through CNTFET channel [13], effect of a dopants or defects in the CNTFET channel [14], use of different types of gates [15], and the use of large diameter CNTs in FET devices [16]. Also there are different types of CNTFETs such as back-gated CNTFET, top-gated CNTFET, partially gated CNTFET, and wrap-around-gated CNTFET, which depend on geometry.

Back-gated type is the earliest way of fabricating CNTFET which involved pre-patterning parallel strips of metal across a silicon dioxide substrate followed by the deposition of CNTs on top of these patterns. This method had many disadvantages: the metal had very little contact to the CNT, and the thickness made it difficult to switch the devices on and off using low voltages. After that, the researchers advanced towards the top-gated CNTFET in which gate is deposited on top of the CNT by evaporation method or by using atomic layer deposition technique. This overcomes the problems of the back-gated CNTFET.

Partially gated CNTFETs are doped uniformly having ohmic contacts. These can be n-type or p-type depending upon the type of doping used. They work in depletion mode. The gate locally depletes the carriers in the CNT and turns off the p-type device with an efficiently positive threshold voltage that approaches the theoretical limit for room temperature operations. Wrap-around-gated CNTFETs are a further improvement upon the top-gate geometry. In this structure, whole of the CNT is gated around in a cylindrical fashion instead of getting just the part of the CNT gated. This technique improves the electrical performance of CNTFET by reducing leakage current and improving the device on/off ratio. The wrapping is partially etched off at the ends of the tube exposing them for contacting with source and drain.

The drain voltage characteristics of conventional FETs and CNTFETs show significantly different behavior. In conventional FETs, characteristics are strongly dependent on the channel material, whereas in nano-FETs, the characteristics are mostly independent of the channel material. In nano-FETs, the drain current as a function of drain voltage is determined by a transmission coefficient of an electron flux flowing from the source to the drain in the Landauer–Butticker way [17–23]. Furthermore, in nano-FETs, the carriers can run ballistically across the channel. In reality, no ideal contact exists; therefore, the characteristics depend on the contacts. However, the gate voltage characteristics show similar behaviors for the conventional FETs and nano-FETs. For both types of FETs, the characteristics are determined by the thickness and the dielectric constant of the insulating layer, which depends upon the surroundings of the channel.

2.6 CNT LOGIC GATES

CNTs are one-dimensional nanostructures with distinctive electrical properties, which make them superior candidates for nanoelectronic applications [24]. One of the key findings is that CNT can replace the channel in FETs and thus named as CNTFET [25,26]. Individual and array of CNTFETs are produced with decent switching characteristics [27]. The next important step in considering the suitability of those devices for electronic applications involves the individual CNTFET integration to form logic gates. These devices are expected to improve the performance while maintaining the transistor stability. Even more thrilling is the possibility of realizing logic functions out of nanotube bundle or a single nanotube. For building such logic circuits, one requires nanotube devices that use holes (p-type CNTFETs) and electrons (n type) as carriers for conduction. Almost all the reported CNTFETs show p-type behavior; that is, they are ON for the negative gate bias. The n-type CNTFET can be obtained by doping with electropositive elements like potassium [28,29]. After the fabrication of individual and array CNTFETs [30], the next important step is to implement integrated circuits from them.

Logic gates are the most basic building blocks of computer logic. The optimum means of making such circuits is to use the complementary devices, which consist of both n and p types. There are many advantages in using complementary devices: they dissipate less power, have large gain, and are easily implemented on integrated circuits. Based on these benefits, complementary logic has ruled silicon technology.

The diameter directly makes an impact on the bandgap and hence influences the Schottky barrier height directly. The I_{OFF} and I_{ON} are inherently related to the Schottky barrier height and thus directly influenced by the diameter of nanotube. Therefore, the I_{DS} versus V_{GS} characteristics can be molded by nanotube diameter. Since small diameter is responsible for the large band gap, higher Schottky barrier results in exponentially lower I_{ON} and I_{OFF}. SWCNTs with large diameter (>2nm) contributes to higher I_{ON} at the cost of poor I_{ON}-to-I_{OFF} ratio and higher I_{OFF}. V_{DD} shows significant impact on I_{OFF} for nanotube Schottky barrier CNTFETs (SB CNTFET). In silicon MOSFETs (nanoscale), the I_{OFF} is affected by applied voltage through drain-induced barrier lowering (DIBL) and therefore shows second-order effect. In SB SWCNTFETs (in contrast to conventional MOSFET), the I_{OFF} is exponentially directly proportional to V_{DD} due to 1D electrostatics of the Schottky barrier.

For further investigations, the design considerations of SB CNTFET are studied. DC analysis for voltage swing and noise margin was performed as the function of supply voltage and tube diameter. Since the voltage swing is the function of I_{ON} and I_{OFF}, the voltage is found to degrade with the applied large value of supply voltage and with diameter of the SWCNT. For ideal inverter, the noise margin should be 0.5 V_{DD}. However, CNT with a diameter greater than 1.5 nm and supply voltage higher than 1V produces higher noise margin. Hence, before designing SB SWCNT, proper care should be taken in choosing the appropriate diameter of nanotube and V_{DD} to achieve the desired voltage swing, I_{ON}, and noise margin. The fabrication of nanotube-based transistors having metal ohmic contacts is very difficult for nanotubes with diameter less than 1.5 nm. This is an emerging field and requires further research consideration to allow SB FET with large I_{ON} and little leakage currents. Furthermore, SWCNT-based MOSFETs are expected to provide better device structure when compared with SB SWCNTFETs for digital electronic applications.

To evaluate the performance-power trade-off, transient analysis has been carried out [31], and it has been found that for the highest throughput, the SWCNT diameter ranges from 1 to 1.5nm. The nanotubes with a diameter less than 1nm are found to have low throughput because of small current drive due to non-ohmic contacts. Alternatively nanotubes with diameter greater than 2 nm provide large I_{ON} but small throughput owing to meager I_{ON}/I_{OFF} ratio. Furthermore, in course of switching between NMOS and PMOS, the delay and short-circuit power increases. Despite large diameter, the SB FET has almost no benefits in accordance with delay, even having high switching currents and are thus practically unwanted for complementary circuits (digital). Thus, nanotubes with a diameter of approximately 1nm and with operating voltage between 0.5 to 0.6V are well appropriate for the digital logic design.

Even though it is recommended to produce highly dense and perfectly ordered SWCNTs, practical processing challenges might lead to non-idealities, such as variation in chirality, orientation, and diameter of SWCNT [32]. Based on previous work, these non-idealities may result in the I_{ON}/I_{OFF} ratio to degrade to below 1,000, which is smaller as compared to today's CMOS-based technology. Large value of I_{ON}/I_{OFF} is recommended for better signal-to-noise ratio in circuit operation. Recently, researchers have obtained chirality of approximately 99% purity [33,34].

2.7 CNT SENSORS

The development of novel methods for sensing is a key theme in current research, and the exceptional properties of CNTs deliver extra potential to explore these areas, e.g., gas sensing. The discovery of oxygen sensitivity to electronic properties of CNTs by Alex Zettl's group appeared as bad news for electronic devices made of CNTs, but it can be explored in gas-sensing applications [35]. Zettl and colleagues in their research work published in 2000 [35] described measurements of thermo electrical power and DC electrical resistance of bundles and thin-film SWCNTs. These samples were placed in a vacuum chamber with heating, cooling, and different gases injection provisions. The resistance measurements were performed at room temperature. The resistance was found to drop by 10%–15% in the presence of air, and the resistance was recovered upon evacuation of chamber. A same effect was perceived when pure oxygen instead of air was used. The work published by Dai and colleagues [36] revealed that the electronic property of individual SWCNTs was very sensitive to ammonia and nitrogen dioxide. In an arrangement used, where titanium and/or gold was connected to individual semiconducting SWNT, using pads as electrodes. The resistance was found to alter by several orders of magnitude when exposed to ammonia and nitrogen dioxide. The increase in conductivity up to three orders of magnitude was witnessed within 10 seconds after the tube was exposed to 200ppm NO_2. With NH_3(1%), the conductivity decreased within 2 minutes by two orders of magnitude. The sensors were operated at room temperature, whereas the conventional sensors for NH_3 and NO_2 consisting of semiconducting metal oxides require operation temperature of about 600°C for high sensitivity. However, the conducting polymer-based sensors have restricted sensitivity. Hence, the SWCNTs are the promising candidates for sensing application.

However, at room temperature, they take many hours to release analyte and heating can be helpful in speeding up. Many other groups explored the application of SWNTs in sensing NH_3, even using functionalized CNTs [37–39]. Functionalized CNTs show greater variation in resistance when exposed to NH_3 as reported by Robert Haddon and colleagues [38]. This effect was ascribed to electron transfer between valance band of semiconducting SWNTs, and attached molecules functionalized tubes for NO_2 sensors have also been used [40]. Alexander Star et al. in 2007 designed SWCNT functionalized sensor with poly (ethylene amine) [41]. NO was initially oxidized to NO_2 and then passed to SWCNT network in a FET device, introducing variations in the conductance of tubes. It has been suggested by authors that the device could be used for asthma diagnosis by NO monitoring in exhale breath.

Dai and co-workers in 2001 first demonstrated that metal particle-decorated nanotubes can also been used in sensing applications and showed that Pd nanoparticle-coated SWCNTs can sense hydrogen from air [42]. Both nanotube bundles and individual SWNTs upon sputter coating with Pd nanoparticles show a decrease in conductivity of 33% and 50%, respectively, when exposed to air with a hydrogen concentration of 400 ppm. The conductivity has been restored when the hydrogen flow is restricted. NASA workers used Pd decorated SWCNTs as methane sensors [43], while another group of researchers fabricated Pd, Pt, Au and Rh decorated SWCNTs for H_2, CH_4, H_2S, and CO_2 sensing [44].

In an exciting study published in 2005, researchers from Sweden and the United States revealed that the electronic properties in metallic SWCNTs show variation when collided with small molecules or inert gas atoms, including Ar, He, Ne, Xe, Kr, and N_2 [45]. These gases are very difficult to discover with existing measurement technologies. CNTs are used as biosensors in many ways. Researchers have shown much interest in replacing macroscopic electrodes (precious metals or glassy carbon) with bulk quantities of CNTs. However, it has been revealed that individual SWCNTs can also be used as sensors. We begin by considering macroscale electrochemical biosensors using nanotube electrodes.

Electrochemical biosensors mostly contain three electrodes: a sink electrode, a sink electrode, and an active electrode. The analyte reacts with active electrode surface, and the ions formed generate a potential that is withdrawn from reference electrode to get a signal. CNTs are promising candidates for electrode selection due to their high conductivities and large surface area. Also, functionalized CNTs have been extensively studied and proven to be better choice with improved electronic transport properties. There are many reports of CNT (electrode)-based devices [46–56] and these have shown superior characteristics as compared to conventional electrode-based devices [57]. There are several methods for making electrodes such as drop coating on glassy carbon [48–51] and mixing nanotubes with binder [46,50]. Nanotubes can also be used for measuring of physical phenomena such as flow rate and pressure, and for "nanobalancing." Daniel Wagner from Weizmann Institute and his colleagues [58–60] have explored the use of nanotubes for pressure sensors. They showed that the strain or stress applied to nanotubes is directly related to induced disorder D* band in the Raman spectrum of SWCNT. This phenomenon can be used to determine the stress/strain in nanotube composite materials. However, if these nanotubes are randomly oriented in matrix form, the interpretation may not be simple. Because when the stress is applied uni-axially to the material, nanotubes may undergo compression.

In order to avoid this discrepancy, nanotubes lying along the polarization direction are selected using polarized Raman spectroscopy as proposed by Wagner and colleagues. With the help of Raman microscopy, the strain distribution in such materials can be mapped. There is a growing interest in gas sensors since the work published in 2000 by Zettl and Dai groups [35,36]. Nanotubes are advantageous over conventional materials having very large surface areas (SWCNTs "all surface), leading to high sensitivity. Furthermore, functionalization or doping of nanotubes with certain catalysts can be used to sense certain gases. An additional advantage of nanotube-based sensors over conventional solid-state sensors (which usually operate above 400°C) is that they can operate at room temperature. Nanotubes for biosensors have also attracted much research interest in gas-sensing applications. More research has already been carried out on electrochemical biosensors (macroscale) involving nanotube electrodes. Thus, nanotube-based sensors offer substantial advantages over conventional electrodes.

2.8 CNT PHOTODETECTORS AND PHOTORESISTORS

The superior mobility due to sp^2-hybridized electronic structure in carbon nanomaterials advocates their use in modern electronic industry. Moreover, the band gap tuning of semiconducting CNTs due to diameter manipulation provides control for modifying optoelectronic properties. Due to these reasons, CNTs are being seen as a potential replacement of conventional semiconducting material (silicon) [61] for optoelectronic and electronic applications.

CNTs with one-dimensional structure and chirality-dependent electronic properties offer a number of eye-catching opportunities for their use in electronic industry: CNTs can be used as channel material in FETs, and also metallic thin-film CNTs are the suitable material as transparent conductor. The exclusive optical property shown by CNTs presents the possibility of developing novel optoelectronic devices. CNTs (semiconducting) provide direct band gap, which results in strong electron–hole pair (excitons) and free electron–hole pair excitation [62,63].

The van Hove singularity existing in the density of states due to 1D nature of nanotubes leads to robust optical emission and absorption, with energy described by CNT chirality [64]. The binding energy of excitons in CNTs is large (around few hundred meV) as compared to ordinary semiconductors as in the case of GaAs, it is less than 10 meV [65] because of strong Coulombic interaction. The improved fluorescence lifetime of 100pS [66] and radio lifetime of 100nS) [67] at room temperature are due to hefty binding energy in one directional excitons. In CNTs, the excitons can be produced electrically and optically as well [68], and the radioactive recombination leads to electroluminescence and photoluminescence in CNTs [69]. The photo-resistivity behavior in SWCNT has been revealed [70]. Recently, the infrared photo-resistivity in individual CNT (semiconducting) has been observed, where the CNT acts as a channel in ambipolar FET [71].

2.9 CNT INTERCONNECTS

An interconnection (a thin film) provides connectivity between two/more nodes designed on a silicon chip made of conducting material. Earlier, aluminum was the most commonly used material because of its good conductivity and stability with silicon dioxide. Furthermore, it forms excellent ohmic contact with silicon. The current density in interconnections increased with the increase in device density. The problem with aluminum was the electromigration with the increase in current density. Later, it was found that copper having high conductivity is more resistant to electromigration than aluminum. Copper has shown the same reliability in IC applications, with five times more current density as compared to aluminum. Due to the large resistance to electromigration, copper is replaced by aluminum, especially in high-density IC devices for improved performance. With advances in IC technology, the device density increased, which forces the cross section of interconnect to be reduced. The decrease in cross section of copper interconnect has resulted in increase in resistivity due to grain boundary scattering and surface roughness which in turn resulted in power dissipation, propagation delay, and electromigration [72,73]. To overcome this problem for future generation interconnects, many solutions are being taken into consideration [72–85]. CNTs are supposed to be the most promising alternative for copper interconnects.

The constant decrease in feature size has resulted in the increase in die size. With the continuous trend of downscaling in electronic industry, it has in turn resulted in upsurge in on-chip interconnects. The interconnects based on the length are categorized as local, semi-global, and global. The local interconnect is the shortest one and connects neighboring nodes on chip. The global interconnect is the longest of all and connects several nodes over the chip like ground line and clock line. The semi global interconnect is the intermediate one in length. The increase in the length of interconnect results in delay in signal propagation. Furthermore, with technology scaling, the propagation delay due to interconnects becomes significant when compared with the delay due to device itself and thus results in performance degradation. According to ITRS [86] for the nanometer-range gate size, the delay in interconnect is mostly due to capacitive and resistive parasitics. For the reduction in resistive portion, a number of replacements to aluminum were proposed like copper which is having better electromigration and electrical resistivity. Furthermore, copper shows higher melting point of 1,357 K as compared to aluminum (993K). Hence, copper offers higher thermal stability. Thus, for these reasons, copper is preferred over other interconnects in integrated circuits.

With the advancement in the VLSI technology, the number of on-chip interconnects increases with time. In order to accommodate the increasing number of interconnects on chip, the cross-sectional dimensions must be reduced, and this results in dimensions in the range of mean free path of electron in case of copper, which is around 40 nm at room temperature. The surface scattering and grain boundary effects are enhanced as the dimensions are reached to the mean free path [87,88]. Consequently, the conductivity of interconnect is decreased. Furthermore, there is increase in the current density with dimension scaling. The above-mentioned effects result in the delay enhancement, and increased interconnect power dissipation due to increase in current density and increase in the operating frequency. The increase in heat dissipation due to power dissipation results in enhanced electro-migration. The limitations posed by scaling in copper interconnects get more severe for future generation integrated circuits. CNTs are proposed to be the best alternative with all their benefits.

The CNTs are single atomic layer of graphene grown in the form of unified cylinder with a diameter of few nanometers. These tubes can be metallic and or semiconducting; in case of interconnects, metallic ones are chosen. The thermal stability of CNTs of as large as 5,800 W/mK, ability to transport current of 10^{14} A/m^2 at 200°C, and Fermi velocity almost the same as metal, proves them as the best choice for future interconnect [89].It is hard to create a good contact with CNT, and the contact imperfections results in high resistance. The resistances of the order of 7–100 KΩ have been reported in the literature [90,91]. Such a large resistance cannot be tolerated; however, the discrepancy can be removed by using CNT bundles instead of individual SWCNT.

A CNT bundle contains a huge number of isolated electrically parallel CNTs. The net resistance between CNT bundles is thus reduced to a considerable amount. Hence, CNT bundles are thus a better choice for interconnects than individual counterparts. The CNTs in the bundle can be either MWCNT or SWCNT. Few of the CNT constituents are semiconducting, whereas others are semiconducting. For interconnects metallic, CNTs play the foremost contribution. SWCNTs are mostly semiconducting, whereas a large fraction of MWCNTs are metallic. DWCNT types of MWCNTs are mostly metallic and are found to be promising candidate for interconnects.

2.10 CNT MEMORIES

A CNTFET can be modified to show a hysteresis behavior in the I_d versus V_G characteristics, which can be used for memory application [92,93]. The silicon flash memory [94], because of its simple double-gated structure and compact size, has been much popular for being used as a memory device. It consists of a control gate and a floating gate. When the floating gate is electronically charged or discharged, the FET threshold voltage changes. In a similar way, an effort is made to build a CNT-based memory device which resembles this flash memory principle. In this case, the CNT is used as a FET channel. For writing "1," the source terminal is grounded, the drain terminal is set to 5 V, and the control gate is biased at 12 V. This puts FET in the pinch-off region which results in the releasing of hot electrons beyond the pinch-off point. These electrons tunnel to the floating gate, and this makes the floating gate to get charged. It is due to this charge that the FET has a higher threshold voltage. In writing "0," the source terminal is put at 12 V, the control gate is biased at 0 V, and the drain terminal is floating. The electrons on the floating gate tunnel to the source terminal which results in the discharging of the floating gate. Thus, the threshold voltage of the FET gets decreased. In reading, V_{CGR} is applied to the control gate. Depending on whether there are electrons on the floating gate ("1") or not ("0"), the FET will have negligible I_D ("1") or finite I_D ("0"). An appreciable threshold modulation was seen in the experimental CNT flash memory operation [95].

Much research is going on to make CNTs possible for memory applications. In the recent past, CNT memory device called NRAM was developed. It is much advanced than the conventional memory devices like DRAM but quite simple in architecture. It has faster read–write speeds than conventional NAND memories and works in CMOS fabrication with no new equipment needed. It can be scaled limitlessly, maybe less than 5 nm in future. It consumes low power, exactly zero in the standby mode. It has the capability to retain memory for greater than 1,000 years at a temperature of 95°C and more than 10 years at a temperature of 300°C. It has high endurance than conventional flash memory used.

2.11 SUMMARY

CNTs are known for their unique electronic properties that make them perfect materials to be used as channels in different types of electronic applications. These unique properties are derived from their 1D character and the peculiar electronic structure of graphite. They have extremely low electrical resistance. They possess high stiffness and axial strength because of the carbon–carbon sp^2 with a measured Young's modulus of 1.4 TPa. It is because of all the remarkable features of CNTs such as ballistic transport and quantum conductance. CNTs are being used in different types of electronic nanodevices. CNTs possess applications in FETs having high efficiency than the conventional MOSFETs. CNTs are also used in sensors and photodetectors with a wide detection range. These are also used in interconnect junctions, logic gates, and memory devices.

Many of the CNT-based electronic devices have been practically demonstrated with great potential applications. CNTs from the beginning were used for making transistors, but now they are used as thin conductive films in the fast-growing touch screen market. CNTs are fabricated as transparent conductive films (TCF), which are highly conductive, transparent, and cost-effective alternative in flexible touch displays. It is a fact that initially the cost of CNTs was high, but in the present world with the technological revolution, it has been coming down as chemical companies build up manufacturing capacity reaching $10/m^2 for film applications. Still, lots of technical challenges are to be overcome for the practical use of CNT-based devices. These include large-scale integration of CNT devices and the control of chirality during CNT synthesis.

REFERENCES

1. S. Iijima, T. Ichihashi, *Nature* 363, 603 (1993).
2. P.J.F. Harris, (Cambridge University Press, New York, 1999).
3. H.W. Zhu, C.L. Xu, D.H. Wu, B.Q. Wei, R. Vajtai, P.M. Ajayan, *Science* 296, 884 (2002)
4. H. Wang, Z. Zu, *Appl. Phys. Lett.* 88, 213111 (2006).
5. N. Hamada, S. Sawada, A. Oshiyama, *Phys. Rev. Lett.* 68, 1579 (1992).
6. P.L. McEuen, *Phys. World* 13 (6), 31–36, (2000).
7. National Physical Laboratory (NPL) of United Kingdom. "Quantised electrical conductance" http://www.npl.co.uk/upload/pdf/ballistic-G0.pdf, 2005.
8. E. Scheer, P. Joyez, D. Esteve, C. Urbina, M.H. Devoret, *Phys. Rev. Lett.* 78, 3535 (1997); *Nature* 394, 154 (1998).
9. R. Saito, G. Dresselhaus, M.S. Dresselhaus, *Physical Properties of Carbon Nanotubes*, Imperial College Press (London, 1998), p. 17–33.
10. T. Hertel, G. Moos, *Phys. Rev. Lett.* 84, 5002 (2000).
11. R.D. Antonov, A.T. Johnson, *Phys. Rev. Lett.* 83, 3274 (1999).
12. H. Dai et al., *J. Phys. Chem. B* 103, 1246 (1999).
13. A. Bachtold et al., *Phys. Rev. Lett.* 84, 6082 (2000).
14. M. Freitag et al., *Phys. Rev. Lett.* 89, 216801 (2002).
15. S. Rosenblatt et al., *Nano Lett.* 2, 869–872 (2002).

16. A. Javey, M. Shim, H. Dai, *Appl. Phys. Lett.* 80, 1064 (2002).
17. W.B. Choi et al., *Appl. Phys. Lett.* 82, 275 (2003).
18. P.G. Collins et al., *Science* 278, 100 (1997).
19. P.G. Collins, H. Bando, A. Zettl, *Nanotechnology* 9, 153 (1998).
20. P.G. Collins et al., *Science* 292, 706 (2001).
21. S. Frank et al., *Science* 280, 1744 (1998).
22. Ph. Poncharal et al., *Science*, 283, 1513 (1999).
23. R. Landauer, *IBM J. Res. Dev.* 1, 223 (1957).
24. Dresselhaus, M., Dresselhaus, G., Avouris, Ph., Eds., *Carbon Nanotubes: Synthesis, Structure Properties and Applications*, Springer-Verlag (Berlin, 2001).
25. S. Tans, A. Verschueren, C. Dekker, *Nature* 393, 49 (1998).
26. R. Martel, T. Schmidt, H.R. Shea, T. Hertel, Ph. Avouris, *Appl. Phys. Lett.* 73, 2447 (1998).
27. P.G. Collins, M.S. Arnold, Ph. Avouris, *Science* 292, 706 (2001).
28. M. Bockrath, J. Hone, A. Zettl, P.L. McEuen, A.G. Rinzler, R.E. Smalley, *Phys. Rev. B* 61, R10606 (2000).
29. C. Zhou, J. Kong, E. Yenilmez, H. Dai, *Science* 290, 1552 (2000).
30. P.G. Collins, M.S. Arnold, Ph. Avouris, *Science* 292 706 (2001).
31. A. Keshavarzi, A. Raychowdhury, J. Kurtin, K. Roy, V. De, *IEEE Trans. Electron Devices*, 35, 2718–2726 (2006).
32. N. Pimparkar, J. Guo, M. Alam, *Digest of IEDM* 1, 120–125 (2005).
33. X. Li, X. Tu, S. Zaric, K. Welsher, W.S. Seo, W. Zhao, H. Dai, *J. Am. Chem. Soc.* 129 (51), 15770–15771 (2007).
34. L. Zhang, S. Zaric, X. Tu, W. Zhao, H. Dai, *J. Am. Chem. Soc.* 130(8), 2686–2691 (2008).
35. P.G. Collins et al., *Science* 287, 1801 (2000).
36. J. Kong et al., *Science*, 287, 622 (2000).
37. X. Feng et al., *J. Amer. Chem. Soc.*127, 10533 (2005).
38. E. Bekyarova et al., *J. Phys. Chem. B.*, 108, 19717 (2004).
39. E. Bekyarova et al., *J. Amer. Chem. Soc.*, 129, 10700 (2007).
40. Q.F. Pengfei et al., *Nano. Lett.* 3, 347 (2003).
41. O. Kuzmych, B.L. Allen, A. Star, *Nanotechnology* 18, 375502 (2007).
42. J. Kong, M.G. Chapline, H.J. Dai, *Adv. Mater.*, 13, 1384 (2001).
43. Y. Lu et al., *Chem. Phys. Lett.* 391, 344 (2004).
44. A. Star et al., *J. Phys. Chem. B.* 110, 21014 (2006).
45. H.E. Romero et al., *Science* 307, 89 (2005).
46. P.J. Britto, K.S.V. Santhanam, P.M. Ajayan, *Bioelectrochem. Bioenerg.* 41, 121 (1996).
47. Z.F. Liu et al., *Langmuir* 16, 3569 (2000).
48. J.X. Wang et al., *Electroanalysis* 14, 225 (2002).
49. C.V. Nguyen et al., *Nano Lett.* 2, 1079 (2002).
50. F. Valentini et al., *Anal. Chem.* 75, 5413 (2003).
51. J. Wang, R.P. Deo, M. Musameh, *Electroanalysis* 15, 1830 (2003).
52. H. Cai et al., *Anal. Bioanal. Chem.* 375, 287 (2003).
53. J.J. Gooding et al., *J. Amer. Chem. Soc.* 125, 9006 (2003).
54. P.G. He, S.N. Li, L.M. Dai, *Synthetic Metals* 154, 17 (2005).
55. G.L. Luque, N.F. Ferreyra, G.A. Rivas, *Microchimica Acta* 152, 277 (2006).
56. A. Arvinte et al., *Electroanalysis* 19, 1455 (2007).
57. J.J. Gooding, *Electrochim. Acta* 50, 3049 (2005).
58. J.R. Wood, H.D. Wagner, *Appl. Phys. Lett.* 76, 2883 (2000).
59. J.R. Wood, Q. Zhao, H.D. Wagner, *Composites A* 32, 391 (2001).
60. Q. Zhao, M.D. Frogley, H.D. Wagner, *Composites Sci. Technol.* 62, 147 (2002).
61. P. Avouris, Z. Chen, V. Perebeinos, *Nat. Nano.* 2, 605–615 (2007).

62. F. Wang, G. Dukovic, L.E. Brus, T.F. Heinz, *Science* 308, 838–841 (2005).
63. M.J. O'Connell, S.M. Bachilo, C.B. Huffman, V.C. Moore, M.S. Strano, E.H. Haroz, K.L. Rialon, P.J. Boul, W.H. Noon, C. Kittrell, J. Ma, R.H. Hauge, R.B. Weisman, R.E. Smalley, *Science* 297, 593–596 (2002).
64. S.M. Bachilo, M.S. Strano, C. Kittrell, R.H. Hauge, R.E. Smalley, R.B. Weisman, *Science* 298, 2361–2366 (2002).
65. G. Grosso, J. Graves, A.T. Hammack, A.A. High, L.V. Butov, M. Hanson, A.C. Gossard, *Nat. Photon.* 3, 577–580 (2009).
66. L. Huang, H.N. Pedrosa, T.D. Krauss, *Phys. Rev. Lett.* 93, 017403 (2004).
67. F. Wang, G. Dukovic, L.E. Brus, T.F. Heinz, *Phys. Rev. Lett.* 92, 177401 (2004).
68. A.A. High, E.E. Novitskaya, L.V. Butov, M. Hanson, A.C. Gossard, *Science* 321, 229–231 (2008).
69. J. Chen, V. Perebeinos, M. Freitag, J. Tsang, Q. Fu, J. Liu, P. Avouris, *Science* 310, 1171-1174 (2005).
70. A. Fujiwara, Y. Matsuoka, H. Suematsu, N. Ogawa, K. Miyano, H. Kataura, Y. Maniwa, S. Suzuki, Y. Achiba, *Jpn. J. Appl. Phys. Part 2* 40 L1229 (2001).
71. M. Freitag, Y. Martin, J.A. Misewich, R. Martel, Ph. Avouris, *Nano Lett.* 3, 1067 (2003).
72. W. Steinhogl, G. Schindler, G. Steinlesberger, M. Tranving, M. Engelhardt, *J. Appl. Phys.* 97, 023706 (2005).
73. A. Naeemi et al., *Electron Device Lett.* 26 (2), 84–86 (2005).
74. C. Schonenberger et al., *Appl. Phys. A* 69, 283–295 (1999).
75. A. Naeemi, J.D. Meindl, *Electron Device Lett.* 26(8), 544–546 (2005).
76. G. Zhang et al., *Proc. Natl. Acad. Sci.* 102(45) 16141–16145 (2005).
77. B.Q. Wei, R. Vajtai, P.M. Ajayan, *Appl. Phys. Lett.* 79(8) 1172–1174 (2001).
78. C. Dong, S. Haruehanroengra, W. Wang, "Exploring Carbon Nanotubes and NiSi Nanowires as On-Chip Interconnections", *Proceedings of ISCAS*, pp. 3510–3513, 2000.
79. A. Gayasen, N. Vijaykrishnan, M.J. Irwin, "Exploring technology alternatives for nano- scale FPGA interconnects", *Proceedings of DAC'05*, pp. 921–926, 2005.
80. Y. Wu et al., *Nature* 430, 61–65 (2004).
81. T. Morimoto et al., *IEEE Trans. Electron Devices* 42, 915–922 (1995).
82. Y. Cui, L.J. Lauhon, M.S.Gudiksen, J. Wang, C.M. Lieber, *Appl. Phys. Lett.* 78, 2214–2216 (2001).
83. Y. Wu et al., *Nano Lett.* 4, 433–436 (2004).
84. K. Toman, *Acta Crystallogr.* 4, 462–464 (1951).
85. B. Meyer et al., *J. Alloys Compounds* 262/263, 235–237 (1997).
86. International Technology Roadmap for Semiconductors, 2007, available: (www.itrs.net/Links/2007ITRS/2007).
87. M.S.Dresselhaus, G. Dresselhaus, P. Avouris, *Carbon Nanotubes: Synthesis, Structure, Properties and Applications*, Springer-Verlag (New York, 2001).
88. W. Wu, K. Maex, "Studies on size effects of copper interconnect lines", *Proceedings of Solid-State and Integrated-Circuit Technology*, Shanghai, China, vol. 1, pp. 416–418, 2001.
89. K. Banerjee, N. Srivastava, "Are carbon nanotubes the future of VLSI interconnections?", *43rd ACM IEEE DAC Conference Proceedings*, San Francisco, CA, pp. 809–814, 2006.
90. Th. Hunger et al., *PRB* 69, 195406 (2004).
91. W. Liang et al., *Nature* 411, 665–669 (2001).
92. M.S. Fuhrer et al., *Science* 288, 494 (2000).
93. M.S. Fuhrer et al., *Nano Lett.* 2, 755 (2002).
94. B.G. Streetman, S. Banerjee, *Solid State Electronic Devices*, Prentice Hall (Upper Saddle River, NJ, 2000).
95. W.B. Choi et al., *Appl. Phys. Lett.* 82, 275 (2003).

3 Electronic Transport Properties of Doped Carbon Nanotube Devices

3.1 INTRODUCTION

Carbon nanotubes (CNTs) realize maximum bulk mobility because of their one-dimensional nature, which severely reduces the phase space for scattering in the material [1,2]. Also CNTs can carry up to 10^9 A/cm^2, due to low scattering, strong chemical bonding, and high thermal conductivity. Furthermore, because of their metallic and semiconducting nature, single-walled carbon nanotubes (SWCNTs) have tremendous potential for electronic device applications. The doping in CNTs brings tremendous changes in electronic properties of the material and therefore results in various electronic device applications. Most of the times elemental metals are used during the synthesis of SWCNTs, which then get mixed up with nanotubes, thus affecting their transport and other properties and making them potential candidates for manufacturing of futuristic nanostructures.

Since the discovery of CNTs by Iijima in 1991, they have attracted and inspired researchers from all branches of science to study their unique nature and morphology. This attraction in CNTs, which consist of a single, rolled sheet of graphene, is because of their unique characteristic features. These extraordinary properties can be attributed to the fact that CNTs are fully carbon-made nanosized in structure having very high aspect ratio due to one dimensionality and they possess a delocalized electronic structure.

The high aspect ratio and delocalized electronic structure of CNTs is responsible for their unique electronic and thermal properties. This also results in heavy confinement in the cylindrical structure which leads to a significant 1D character of the electronic distribution and the phonon spectrum. The 1D character of CNTs results in very weak electron–phonon interaction. Therefore, ballistic charge transport is seen in these materials even at room temperature. Also high values of thermal conductivity in the range of 10^3W/km which is in the vicinity of diamond is achieved. Furthermore, the wrapped structure and delocalized electron system makes CNTs almost insensitive to lattice defects. On the basis of these properties, CNTs are considered to be ideal 1D structures for fundamental research and development. CNTs are actively being studied for their use as field-effect transistors (FETs) [3–5], diodes, logic gates, oscillators, interconnects, and memory elements [6].

CNTs are fully made up of carbon atoms having high surface exposed morphology. This type of structural formation attracts researchers from different fields such as medicine, chemical sciences, bio-sciences, and physical sciences. It is due to the catenation of carbon and its abundance in nature that it is one of the most researched materials in the scientific community. Thus, as a result of being all carbon-based structure with maximum exposed surface, CNTs are considered as excellent building blocks for nanosized materials. Due to these remarkable properties, CNTs are being intensely studied for their application in biosensors, gas detection, drug delivery, and gene therapy [7–13]. Moreover any alteration in the delocalized electronic structure results in changing of electronic properties of CNTs. Therefore by introducing doping in the CNT channel, electronic properties can be tuned in different ways. This tuning of CNT properties opens up a vast area of research across the globe.

The dopants can be inserted at different positions which alter the electronic transport properties of CNTs [14]. Due to the unique structure of CNTs, we can use different types of doping approaches like doping along the channel which greatly changes the electronic properties of the CNTs [15]. These features make CNTs interesting for comparative studies of different dopant locations and concentrations to alter their electronic transport properties in a co-doped system. A lot of work has been done in the past to study the theoretical aspects of doped CNT systems [16–19]. A large number of experimental and theoretical methods have been used in nanostructure-based devices to introduce other types of exciting transport properties like negative differential resistance (NDR) [20].

CNTs are chemically stable and possess perfect structure which enables it to show unique carrier mobility at high gate fields without getting affected by processing and roughness scattering as observed in the conventional semiconductor channel [21–23]. CNTs are of much importance in devices due to their 1D nature of density of states (DOS) which greatly reduces the scattering region area giving the high on-current in semiconducting CNT field-effect transistors (CNTFETs) [24]. The gate control is also improved due to the absence of dangling bonds on surface in one-dimensional CNTs. The one-dimensional character also results in suppression of short channel effects of CNTFET devices.

CNTFETs have been fabricated using semiconducting CNTs, which show far more superior electrical characteristics over silicon-based metal oxide semiconducting field-effect transistors (MOSFETs) [25]. The promising characteristics of single CNTFET have led to initial attempts at integration of several CNTFETs into useful circuits which can be used in a large number of electronic operations or function as memories or sensors [26]. The CNT logic gates have been, in most cases, based on a complementary technology analogous to silicon CMOS, which is important as it may ease integration of CNTs onto this well-established technology. It is to mention that various types of CNTFETs are discussed by different researchers like CNTFETs with gates at the back of the device or on the top of the device, and CNTFETs with contact-dependent electrodes.

The properties of CNTs are altered by introducing dopant atoms which causes variation in the conductivity depending upon the dopant location and the type of doping used. The doped CNT channels in FETs result in amazing characteristics and as such huge research is going on to understand the mechanisms behind the CNT transistor operations in order to enhance their performance [20]. Many attempts are being made to explore the unique properties like NDR behavior in CNTFET. Lots of software packages like Atomistic ToolKit (ATK) are being used to study the various transport properties using first principle approach [27].

ATK software is one of the most popular commercial software for atomic-scale modeling and simulation of nano-based structures. It provides easy interface for modeling and simulation of different electrical and transport properties in nanodevices. The software uses enhanced scripting language with highly accessible graphical user interface. ATK provides modules for analysis using first principle approach like density functional theory (DFT) and high-speed semi-empirical approaches of various classical potentials.

This chapter deals with doping methods and techniques, transport properties of two-probe and three-probe devices, and CNT bio-molecule sensors.

3.2 DOPING METHODS AND TECHNIQUES

The unique structural design of nanotubes makes them ideal for doping by a wide variety of possible approaches and thereby changing their physical and electronic properties. It is due to the molecular nature of CNTs; doping them can be easy by adopting semiconductor industry methods like substituting carbon atoms by impurities such as boron, nitrogen, and silicon. Many doping approaches are being used in order to explore the various doping-dependent characteristics of CNT devices.

The electronic properties can be changed by introducing dopants at proper positions in the CNT channel. Because of uniqueness in CNT structure, different ways of doping methods are used like replacing atoms along the scattering channel, which in turn results in alteration of electronic properties. These vast numbers of doping approaches make the comparative study of the dopant positions and concentrations very exciting. Doping can be done by inserting the dopant inside the CNT channel, known as endohedral doping or encapsulation [28–35], by attaching the dopant on the outside surface of the channel known as exohedral doping or intercalation [36,37] or by replacing the carbon atom with the dopant atom on the surface known as substitutional doping [38–42]. CNTs can also be doped by either electron donors or electron acceptors [43].

The endohedral doping includes the capillarity action in CNTs which made it possible to capture atoms or molecules inside CNT channel. This encapsulation by CNTs has been a very interesting field of research and development. The discovery of C_{60} SWCNTs known as peapods has led to much advancement in CNT research. The controlled filling by C_{60} peapods could lead to development of double-walled carbon nanotubes (DWCNTs) after proper heat treatment. The dynamics of water in CNT channels also adds to the research interest in fluid dynamics of biological systems like proton pump in cellular transport. Furthermore, the capture water molecules in the CNT channel open up new area of studying chemical reactions inside the channels leading to a unique nanoreactor.

The exohedral doping includes the fact that SWCNTs act as host sites in which electronic properties are altered by doping the channel with electron-donating or electron-accepting impurities and small molecules which get adsorbed at the interstitial sites. This doping increases the free charge carrier density, thus enhancing the electrical and thermal conductivity in the CNT bundles. After reaction with host materials, the dopants are intercalated in the inter shell spaces of the multiwalled nanotubes, and in the case of SWCNTs either in between the individual tubes or inside the tubes.

The replacement of carbon atoms in the CNT structure by electron rich or deficient atoms introduces additional states in the DOS of the CNT system. Electron donating or accepting nature of the states depends upon the local bonding pattern of the dopant atoms used. Substitutional doping of B and N in CNTs will introduce strongly localized electronic features in the valence or conduction bands, respectively, and will enhance the number of electronic states at the Fermi level depending on the location and concentration of dopants.

Some of the doping techniques are listed as follows:

1. Functionalization of oxidized SWCNTs with carboxylic acid groups forming amides and esters [44].
2. Functionalization of CNTs with nitrenes, carbenes, and radicals [45–53].
3. Functionalization of CNTs with alkali metals [54,55].
4. Functionalization of CNTs with lithium alkyls [56–58].
5. Azomethine cycloaddition process with CNTs [59,60].
6. Electrochemical modification of CNTs [61].
7. Functionalization of nitrogen-doped CNTs with various chemical groups [62].
8. Fluorination of CNTs [63–68].
9. Endohedral or encapsulated doping [28].
10. Exohedral or intercalated doping [36].
11. Substitutional doping of CNTs [38].

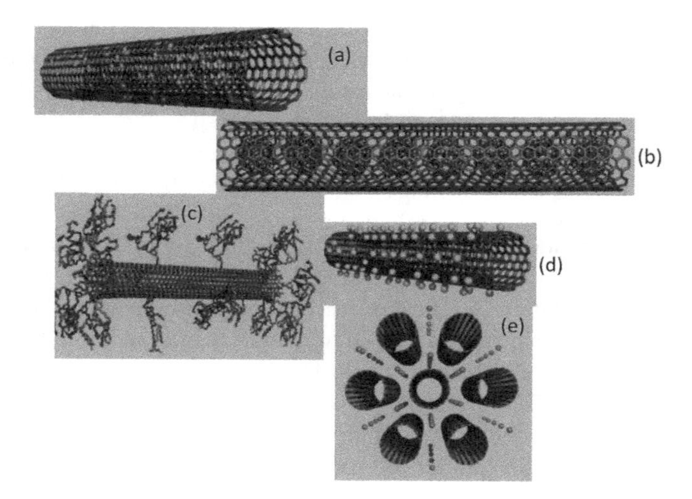

FIGURE 3.1 Various doping approaches in CNTs. (a) substitutional doping, (b) CNT filled with fullerenes called peapods, (c) functionalized CNT, (d) fluorinated CNT, and (e) CNT bundles intercalated with atoms or ions.

Figure 3.1 shows different approaches of chemical modifications in CNTs. Attaching of functional groups on the CNT surface has revolutionized the doping process in CNTs by taking advantage of their extraordinary physical properties, such as high surface area, electric conductivity, and tensile strength [69–75]. CNTs are highly inert which enables them to react with only highly reactive chemicals in order to form covalent bonds with the sidewalls. The inert nature is because of homogeneous delocalized π-system. On the scale of inertness with respect to all the sp^2 hybridized allotropes of carbon, CNTs are placed in between the fullerenes and graphite. Functionalization of CNTs makes them promising candidates for a number of potential applications such as use in reinforced and conductive plastics, sensor elements, photovoltaic material, and scanning probe microscope.

3.3 TRANSPORT PROPERTIES OF TWO-PROBE CNT DEVICES

CNTs offer good electrical conductivity, thermal conductivity, and high chemical, thermal, and mechanical stability, which make them one of the most promising subjects of nanoelectronics. They have already explained themselves quite well, as perfect candidates of electronic systems as electron field emitters, supercapacitors, solar cells, nanoelectromechanical systems, and sensors. CNTs act either as building blocks or form an active material in a large number of nanodevices, ranging from FETs, bio-molecule sensors, optoelectronic devices, and nanoelectromechanical systems (NEMS). One of the first electronic nanodevices manufactured using CNT was the normal FET with a backgate. The design used at that time is still the basic step for any type of modern CNT-based nanodevices. The common CNTFET device design consists of a CNT lying on the top of two electrodes which are placed on the dielectric surface with a back gate. In newer designs of CNTFET, the electrodes are placed on top of the CNTs to improve contact stability with the same type of back gate used. The developments are going on to increase gate efficiency by employing different techniques of attaching gates with the CNT channel. One of the methods include using topgate, in which gate is evaporated on top of the device.

The main contribution to the conductivity of CNTs is made by electrons. Therefore, the conductivity of CNTs is determined by the mechanism of electron mobility. In a defect less SWCNT at low temperatures, electrons pass through the nanotube without scattering, which corresponds to the quantum ballistic mechanism of conductivity. Electrical transport in CNTs has been investigated experimentally and theoretically by many research groups across the globe. Delocalized π-electron system is responsible for conduction in CNTs. SWCNTs are ballistic conductors at room temperature with two and one spin-degenerate conducting channel in metallic and semiconducting types, respectively. The channels belong to the first π- and π*-band of the delocalized π-electron system. Thus, at ambient conditions, without external excitation, SWCNTs have only two spin-degenerate conducting channels available for charge transport, and each channel can carry the conductance quanta G_0, given as [76]

$$G_0 = \frac{2e^2}{h} \tag{3.1}$$

where "e" is the electron charge, and "h" is the Planck constant.

The energy gap in semiconducting CNTs is inversely dependent on the nanotube diameter [76]. The weak scattering cross section and negligible electron–phonon interactions result in ballistic conduction of charge carriers in CNTs. It has been shown that the mean free path of metallic CNTs lies in the range of few micrometers using femtosecond time-resolved photoemission calculations [77]. Such a large mean free path is much above the normal lengths of CNT devices, in which the length of conducting channel falls in the sub-micron regime. Thus, CNTs can be considered for use in low power consumption electronic devices.

3.4 EFFECTS OF DOPING ON ELECTRONIC TRANSPORT PROPERTIES OF TWO-PROBE CNT SYSTEMS

CNT device consists of left and right electrodes and the central scattering region as shown in Figure 3.2. All of the three parts can be doped for exploring different properties, but here we concentrate only on doping of the central scattering channel [14]. Different elements such as boron, nitrogen, chromium, indium, and arsenic are used as dopants.

Figure 3.3 shows the CNT model with two electrodes and the scattering region doped with two atoms. The bond length between the carbon atoms is 1.42 Å, and the two electrodes are 7.1 Å in length. In order to minimize the scattering loss, 10% of

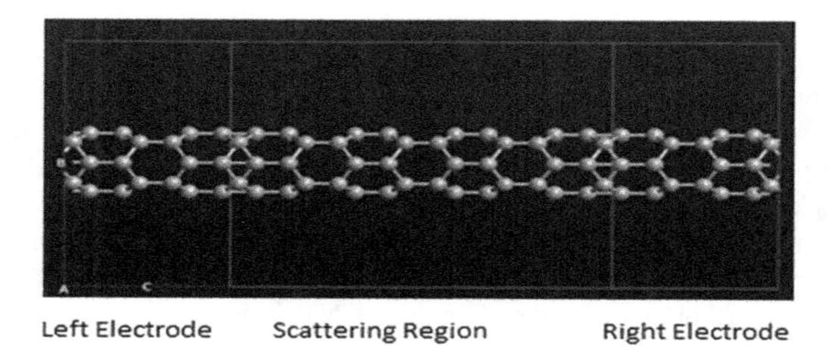

FIGURE 3.2 SWCNT device showing scattering region, left electrode, and right electrode.

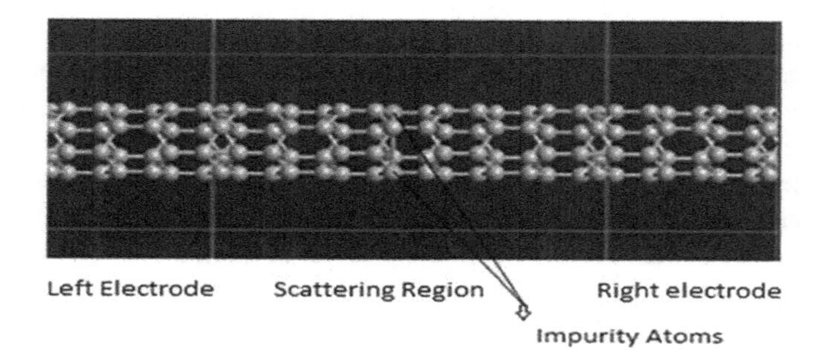

FIGURE 3.3 Illustration of doped CNT two-probe device.

the electrode length is considered within the scattering region. The voltage is applied across the two electrodes of this two-probe CNT system to observe the electronic properties under low bias and to study the changes caused due to the doping of central scattering region.

To study the effect of dopants on the electronic properties of the two-probe CNT system, the proposed model is simulated in ATK software package using first principle approach. The model is simulated under different applied voltages varying in the range of 0–1 V. The scattering region is substitutionally replaced with dopant atoms of different elements such as chromium, arsenic, tellurium, and antimony. In simulation models, inelastic scattering processes are assumed to be negligible making ballistic transport possible across the device. The simulations for calculating transport properties are based on NEGF formalism with single particle approach.

Table 3.1 gives the conductance values obtained from various CNT models doped with different dopant atoms. The maximum conductance is observed in arsenic doped model, and the minimum conductance in chromium doped two-probe CNT devices which also show a remarkable property of NDR behavior. NDR may be defined as a region in the I-V curve where an increase in device voltage results in a decrease of electric across the device, which is totally different from a common resistor where an increase in applied voltage across the device results in proportional increase in current, thus following ohm's law and showing positive resistance. In case of a positive resistance, power is consumed, whereas in case of negative resistance power is produced. Thus, it results in amplification of the electrical signal.

It is also clear that tellurium-doped CNT model shows a conductance of 60921.8 nS followed by terracing slopes in the I-V curve. Both antimony and arsenic doped models have almost same increase in conductance, but the latter one has sustained maximum conductivity over a wide range of applied voltages.

TABLE 3.1
Values of Conductance for Two-Probe CNT System at Different Bias Voltages

	Conductance (nS)			
Applied Bias (V)	**Tellurium**	**Antimony**	**Arsenic**	**Chromium**
0	0	0	0	0
0.1	60,921.8	94,320.6	95,528.3	84,072.5
0.2	42,600.65	60,461	61,543	54,958.5
0.3	49,428.33	70,652.33	72,068	55,572.66
0.4	38,199.5	56,411.25	57,604.25	21,693.8
0.5	42,502.2	64,007	65,079.8	22,613
0.6	39,205	59,223	60,572	11,714.3
0.7	45,246.71	65,970.57	68,093.28	9,582.94
0.8	42,934.375	65,546.75	53,020	8,465.81
0.9	49,864.77	71,211.11	65,274.2	9,003.18
1.0	52,304.5	69,622.7	69,750.7	10,408

The maximum values of conductance are observed in antimony-and arsenic-doped CNT devices. Arsenic doping in CNT enhances conductivity because of the maximum modulation of Fermi level as compared to other doped CNT devices. On the other hand, chromium doping in CNTs shows multiple NDR regions with increase in the applied bias voltage across the device.

Thus, the dopant atoms alter the band gap of the CNT channel showing different characteristics based on the type of dopant used. Also the conductance of each modeled device is different with changing bias voltages. It is also concluded that doping process modifies the electrical transport properties of CNT devices with respect to the dopant concentration, type, and position in the scattering channel. Modification of the Fermi level in doped CNT channels causes changes in the band structure also which affects the transport properties of doped CNT devices. The changes in conductivity of doped CNTs are useful for many applications in the electronic devices such as design of CNT-based adders, multipliers, and other architectures. For example, chromium-doped CNT device shows NDR effect which is suitable for the design of amplifiers, oscillators, etc.

3.5 NEGATIVE DIFFERENTIAL RESISTANCE (NDR) IN CNTFETs AND CNTMOSFETs

Negative differential resistance (NDR) is described as an increase followed by a decrease in current with increase in applied bias voltage. Different research groups have been working on this feature to explore the mechanism behind NDR effect, the idea of which would greatly affect the field of molecular electronics [78,79]. Different explanations have been used to understand the NDR behavior in metal–molecule interfaces [80–84]. In weakly coupled junctions, it is stated that the closely packed DOS of the tip apex atom is responsible for showing this behavior [85–87]. On the other hand, it has been recently shown that the matching of symmetry of local orbitals between the tip and the molecule is the cause of the NDR behavior [88]. Also, the charging and conformational changes due to the applied bias range in the molecule are thought as a viable mechanism for introducing NDR behavior in the system [89].

Great interest has been shown by researchers all over the world for inducing NDR behavior in decreasingly smaller devices in order to reach high device integration densities in electronic industry. Another fact is that the capacitance in the NDR regime is proportional to its size, which restricts the accessibility of the device characteristics for large values. In the recent past, there have emerged CNT devices like CNTFETs that exhibit NDR in their electrical characteristics induced due to different mechanisms like by the creation of defects, introduction of chemical doping, heterojunctions, and quantum dots in between the metallic interfaces and the multiwalled CNTs [79–83]. These studies prove CNTs to be used as NDR components in futuristic electronic applications. Figure 3.4 shows the CNTFET doped with different concentrations of boron and nitrogen atoms to induce the NDR behavior.

The drain–source current is calculated as a function of drain–source voltage in the range of 0–2 V. The I-V characteristics of the CNTFET device showing NDR behavior is given in Figure 3.5. From this figure, it is clear that the CNT-based FET shows NDR behavior. The possible mechanism is the band-to-band tunneling (BTBT) caused by the proper distribution of acceptor and donor states. The doping

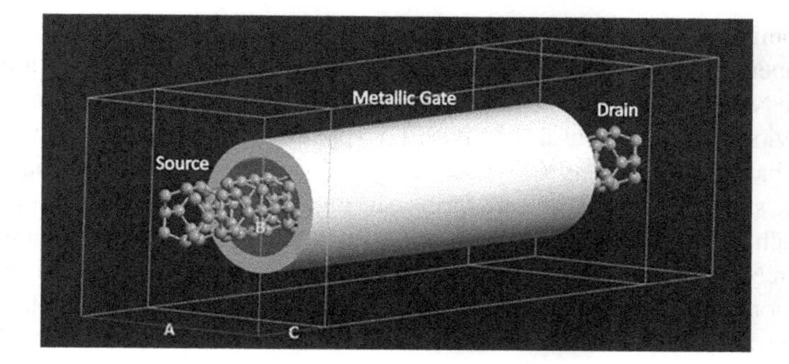

FIGURE 3.4 All around gated CNTFET.

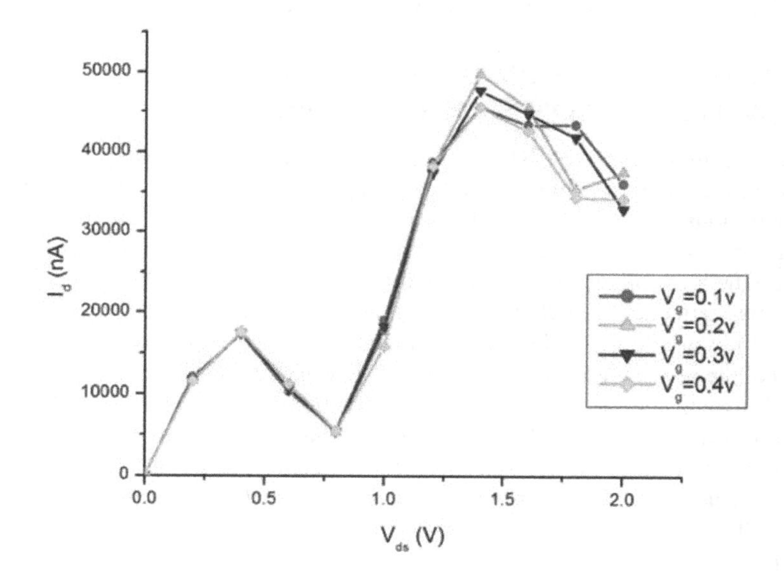

FIGURE 3.5 Output characteristics of CNTFET keeping gate voltage constant at 0.1, 0.2, 0.3, and 0.4V.

of boron and nitrogen atoms introduces p-type and n-type character, respectively, in the CNT channel resulting in creation of donor–acceptor-like states. As the field is applied along the CNT channel, it shifts the valence and conduction bands with respect to the energy gap of the semiconductor, thus making BTBT possible. When semiconducting CNT channel is used in combination with boron–nitrogen doping, shifting of bands takes place with respect to Fermi level, thus producing NDR effect in the doped CNTFET device. This characteristic feature in CNT device is very useful in nanoelectronic applications such as digital logic circuits, oscillators, amplifiers, and memory devices.

The NDR behavior in conventional MOSFETs is also observed. In short-channel MOSFETs, as the quantum confinement increases with short channels, NDR becomes more prominent. In nanoribbon (NR) MOSFETs, the NDR is observed more easily. This happens because of lateral confinement in the NR structure [90]. In the recent years, the NDR has also been observed in MoS_2 armchair NR, showing semiconducting behavior, making it useful for MOSFET applications. This will lead to enhanced memory-based applications of MOS devices. Two-dimensional materials such as graphene, silicene, and monolayer MoS_2 can undergo a certain amount of deformations which are attributed to different causes such as thermodynamic instabilities and strains. In NRs, the most deformations include twisting and wrapping [91–93]. These deformations involve a prominent effect on the functioning of MoS_2 channel with regard to transport and NDR in MOSFET device. The NDR behavior in MOSFETs is useful in the downscaling of many electronic elements such as memory circuits, amplifiers, and RF oscillators. Moreover, the development in studying the NDR behavior in MOSFETs will revolutionize the mechanical sensor application of FETs.

3.6 NDR IN CHROMIUM-DOPED SINGLE-WALLED CARBON NANOTUBE DEVICES

In order to understand the NDR effect in chromium-doped SWCNT system, the central scattering region of CNT device is doped with chromium atoms of different concentrations in a pyridine-like shape as depicted in Figure 3.6. The doping is done in between the two on tube electrodes in order to predict the transport properties through the junction under varying conditions of bias voltage [94].

In this particular research, real-time scattering is done using the extended Huckel theory. First of all, the models are optimized and the structures are fully relaxed to get rid of the residual forces on each atom. The value for density mesh cutoff is taken as 10 Hartree. The maximum range of interaction is set as 10 Å with the K-point sampling values put as $1 \times 1 \times 100$ for fast and accurate calculations. Fast Fourier 2D method is used as the Poisson solver for boundary conditions. The simulation is done with electrode temperature maintained at 300 and 400 K separately for all the models. Then, the I-V curves are obtained and analyzed.

Table 3.2 shows the peak to valley current ratios of the observed I-V characteristics of three CNT structures doped with different concentrations of chromium atoms. In two-atom-doped CNT-based model, NDR peak can be visualized twice in the current–voltage curve as shown in Figure 3.7a. First peak appears at a bias voltage of 0.3 V, and the valley point occurs at 0.8 V. The peak-to-valley ratio is calculated as 5.70. Similarly, another peak is observed at a bias voltage of 1.6 V, and

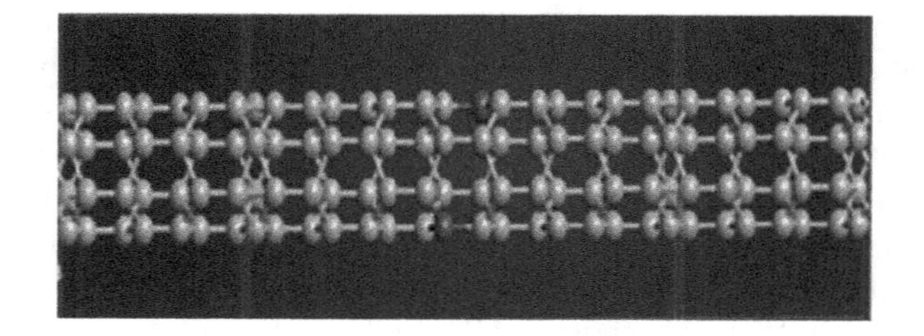

FIGURE 3.6 Chromium-doped SWCNT device.

valley occurs at a bias voltage of 2.3 V. The peak-to-valley ratio is calculated as 2.4 which is less than the first one.

In another model doped with four atoms of chromium, SWCNT device shows two NDR regions as shown in Figure 3.7b. First peak appears at a bias voltage of 1.36 V, and the valley point occurs at 2.0 V. The peak-to-valley ratio is calculated as 1.68. Similarly, another peak is observed at a bias voltage of 2.76 V, and valley occurs at a bias voltage of 3.35 V with the peak-to-valley ratio calculated as 1.83. With further increase in the concentration of doping atoms, multiple NDR regions are observed in the I-V characteristics of the device as shown in Figure 3.7c. The results thus show that NDR effect strongly depends upon the concentration of chromium dopant atoms. In the eight-atom-doped model, first peak is seen at an applied voltage of 0.3 V, and valley occurs at around 0.6 V with the peak-to-valley ratio calculated as 2.64. The second peak occurs at a voltage of 1.41 V and valley point at a voltage of 1.95 V with the peak-to-valley ratio calculated as 1.6. The third peak occurs at a voltage of 2.52 V, and valley is observed at a voltage of 3.76 V with the peak-to-valley

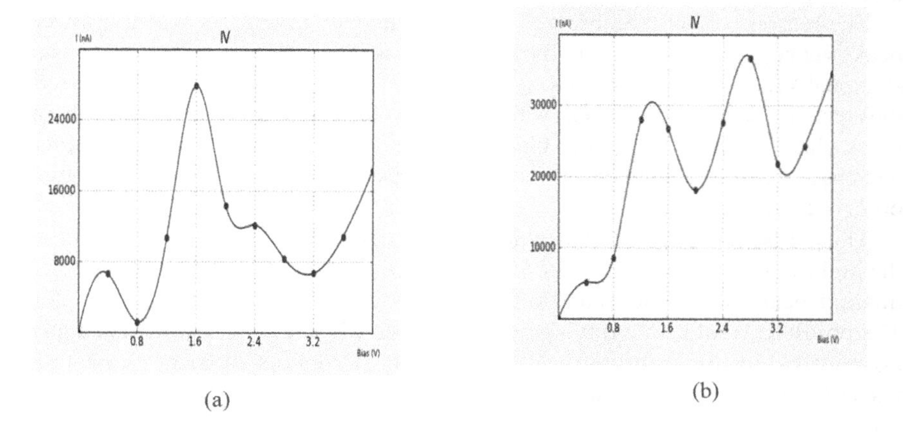

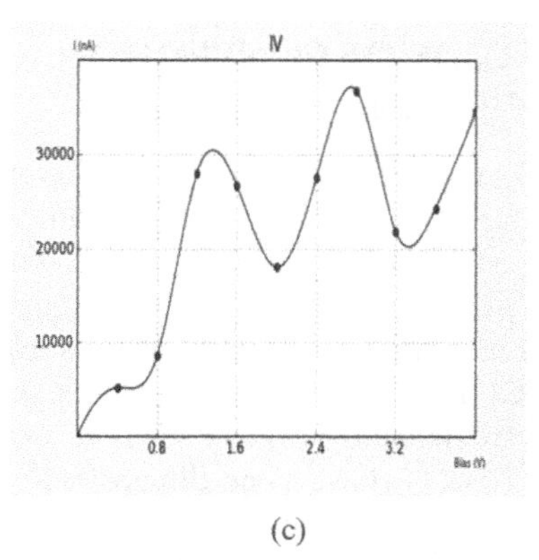

FIGURE 3.7 Current–voltage curves of (a) two-atom, (b) four-atom, and (c) eight-atom chromium-doped SWCNT devices.

TABLE 3.2
Peak-to-Valley Current Ratios of the Three Proposed Models at 300 K

PVR of Two-Atom-Doped SWCNT Device			PVR of Four-Atom-Doped SWCNT Device			PVR of Eight-Atom-Doped SWCNT Device			
Ip	6,850	27,600	*Ip*	30,400	37,120	*Ip*	3,175	24,800	37,600
Iv	1,200	11,500	*Iv*	18,000	20,200	*Iv*	1,200	15,500	20,000
PVR	5.70	2.4	PVR	1.68	1.83	PVR	2.64	1.6	1.88

ratio calculated as 1.88. From Table 3.2, it is also clear that two-atom-doped SWCNT device showed height peak-to-valley current ratio of 5.70 followed by eight-atom-doped SWCNT device of 2.64. These remarkable characteristics from the observed results show that tunneling current and the NDR regions increase with the increase in doping concentration.

The reason behind the observed NDR regions lies in the fact that transmission peaks get suppressed as seen in the simulated results. This suppression of transmission peaks results in less probability for electrons to flow across the two electrodes, thus causing decrease in current with increasing bias voltage. Also the shift in DOS takes place significantly as the bias voltage is increased which in turn results in decrease of overlap of states around Fermi, thus forming multiple NDR regions as observed in the I-V curve.

Thus, it is concluded that chromium-doped CNT device shows NDR effect with the highest peak-to-valley ratio of 5.70. The NDR is due to the suppression of transmission peaks which depends on the concentration of chromium dopant atoms. The observed results are promising for doping of CNTs and their potential applications such as logic circuits, frequency multipliers, oscillators, memory elements, and fast switching electrical devices.

3.7 COMPARATIVE STUDY OF CONVENTIONAL AND ELECTRICAL DOPING IN CNT DEVICES

Since the discovery of CNTs in 1991 by Sumio Iijima, continuous studies have been carried out to understand their electrostatic behavior, transport properties, and performance analysis for their use in high computing systems which are economical and consume less power. CNTs are attributed with a wide range of potential applications because of their small size, unique transport characteristics, and one-dimensional nature. For the growing need of compact and efficient design, CNTs are expected to be used in different electronic nanodevices like FETs to overcome the size limitations of silicon FETs [95–97]. CNTs have a one-dimensional DOS due to which scattering phase space gets reduced resulting in suppression of short channel effects and high on-current, besides the absence of surface dangling bonds provides enhanced gate control with negligible leakage. The first CNTFETs were fabricated, and their performance was analyzed in the year 1998.

CNT properties like channel conductivity can be easily modified by different types of doping techniques, which on using as a channel in different devices like FETs provide extraordinary results, thus providing a huge interest in better CNT transistor operations and enhanced performance [98–101]. The extraordinary morphology of CNTs provides different possible approaches to doping and altering their physical and electronic properties. Different types of conventional doping methods are used in CNTs like functionalization by carboxylic acid groups, nitrenes, carbenes, radicals, alkali metals, and lithium alkyls, electrochemical modifications, fluorination techniques, and substitutional replacement of atoms. But these conventional doping methods in CNTs cause various disadvantages such as change in electrochemical properties of the material, heat generation, instability of the doped structure, interatomic reactions between CNT and the dopants, and compatibility problems. Therefore to overcome these issues, here in this study we have introduced the concept of electrical doping in CNTs in which no external impurities or atoms are used.

Electrical doping is proving to be one of the promising techniques in the fabrication of fault-free devices [102–104]. This doping technique is different from the conventional doping process. In this procedure, conventional p-type and n-type dopants are not required; instead, carriers are injected by applying the electrically same and opposite charges at the two ends of the device creating a potential drop, thus enhancing the transport properties of the device. One of the remarkable characteristics of electrical doping is that by increasing carrier injection and reducing the operating voltage, the efficiency of the device can be improved. As compared to the conventional doping methods, the electrical doping process facilitates carrier injection, electrical resistance, forward voltage, and carrier recombination without breaking the interfacing layer. Also, the formation of doping induced bands is prevented in the electrical doping process, enabling the use of high doping concentrations. The results reveal that the maximum current of about 100 μA is obtained in case of electrical doping as compared to the conventional doping of CNT with group 3A and group 5A elements of the periodic table.

Electrical doping at the two ends of the device creates a potential drop due to the formation of p-type and n-type regions. The conduction channels move towards the Fermi level shifting the energy gap towards the lower energy level, thus contributing to the carrier transmission. The electrical doping method produces higher conductivity as compared to the conventional doping approach. The conductivity of the device channel depends on the quantum transmission which in turn depends on electrical doping. This makes conductivity dependent on electrical doping. Thus, a large number of available energy states lead to high channel conductivity. The electrical doping at the two terminals of the device creates a potential drop acting as a driving force for channel conductivity. As the electrical doping concentration is increased, the potential along the channel also increases, thus increasing the channel conductivity.

Conventional doping works by introducing new electronic states near the Fermi level and also lowers the orbital localization which enhances the tunneling between the two electrodes. Thus, the increase in conductance takes place along the device channel. Conventional doping is the widely used technique in CNTs and also enhances the conductivity but is not so effective as compared to the electrical doping approach. Conventional doping inserts various structural defects due to the interaction between the dopants and the host atoms altering the DOS. The weak interaction between the dopant atoms and host atoms results in compatibility issues, thus affecting the stability of the device. In order to overcome these problems, the easier method of electrical doping has been introduced which gives much more efficient results than the conventional doping.

3.8 TRANSPORT PROPERTIES OF CNT BIO-MOLECULE SENSORS

The molecule-based sensor market has been estimated to reach billions of dollars in upcoming decade or so. Potential applications of these molecular-level devices are checking and controlling of environmental pollution, enhancing diagnostic methods medical purposes, and producing small, power-efficient, reliable, and fast instruments for industrial processes and application in warfare and security purposes. For all these advantages to be more effective, the sensors should be improvized in sensitivity, efficiency, selectivity, and stability to make them far more useful than other sensors which are available commercially. In order to achieve this goal, the researchers working on CNTs have proposed various methods for sensor applications, which are based on making changes in a particular property of a system like chemi-resistors and back-gated FETs are based upon the changes in CNT conduction, magnetic resonant sensors detect the change in frequency due to mass loading effect, and optical sensors which are sensitive to scattering of light caused by the different processes.

The sensor-based uses of CNTs are continuously being researched upon all over the world. CNTs are now thought to act as new types of adsorbent material, thus holding an important position for carbon-based detector material. CNTs have large surface area which is chemically inactive providing feasible conditions for physical adsorption, thus making a large number of adsorption sites available for different kinds of molecules. However, varying charge distribution due to the charge transfer and change in adsorption energy because of gas morphology contributing together in the adsorption process provide the qualitative and quantitative mechanisms for changes in conductivity during gas adsorption experiment of the CNT detectors, thus making gas detection process possible.

CNTs on the minimum possess two characters which are required for highly sensitive biosensors; that is, size should be compatible with different types of biomolecules and cells, and charge transport which is totally on the CNT surface [105]. CNT sensor-based applications are done through a sequence of simulations using different molecules as target materials which are to be sensed. Molecular dynamics (MD) methods are also used in these simulations with different types of reaction potentials. Investigation is done on the effects of hydrophobic and Van der Waals forces on the adsorption of target molecules by CNTs. Also the relative positions between the CNT and the target molecule are determined. These parameters are then analyzed using different methods like DFT or NEGF in various simulation packages, the main purpose of which is to visualize the changes in the DOS of CNT caused by the target molecule adsorption.

These CNT-based sensors will revolutionize the medical field by enabling the concerned researchers with a fast and accurate diagnostic tool. This can be done by obtaining information about the composition of different chemical substances in exhaled air or excretory products from the body. By this technique, different types of metabolic syndromes, skin diseases, and infections can be diagnosed. Furthermore, these sensors help in providing more reliable and accurate real-time patient information during surgical procedures.

The advancements in synthesis of nanodevices further inspire their use in bioengineering field. An acoustic sensor based on CNTs which is very sensitive to diminishing sounds can be used as a nanostethoscope to listen very faint sounds in nature. It can detect and monitor the sounds caused by the presence or movement of cancerous cells which are much active than other normal cells present in the body. A new type of surface called "smart skin" can be made by combining IR detector and sound sensors based on MWCNTs which is sensitive to light, changes in temperature, and changes in frequency. CNT-based actuators can function at very small voltages and in turn result in large strains, thus providing a possible future of CNTs being used as artificial muscles which will be a boon in the medical field.

3.9 SUMMARY

In summary, it is concluded that the there is a wide range of scope for CNT devices. CNTs can be used a channel material between two electrodes which shows excellent characteristics. There exist a number of approaches for CNT doping such as substitution doping, exo-doping, and endo-doping. It is to mention that the nanotubes are not currently a homogenous material. On the basis of their single or multiple walls, their properties change drastically. As per this criterion, SWCNTs can be metallic or semiconducting which depends upon the chirality of the nanotube. It is a fact that the nanotube samples normally contain a distribution of tube diameters, number of tube walls, and mixture of single-walled and multi-walled nanotubes. And therefore, their properties are not the same. Thus, it is of great importance that care should be taken to determine the measurements on individual CNTs.

Doping of CNTs has revolutionized the field of electronic devices and has removed the limitations of conventional semiconducting materials. Furthermore, the role of CNTs in nanoelectronics is likely to change drastically with the development of technology over growth, manipulation, and functionalization of the channels. Also the development in understanding and control over designing processes would revolutionize the future of CNT-based devices. Research is going on very rapidly in many similar fields in order to make possible the controlled growth of CNTs with specified parameters and helicity onto chip surfaces. This will lead to the new world of microprocessors with small component sizes, faster processing speeds, improved thermal management, and very low leakage. CNT-based devices offer the sense of making novel futuristic device designs for large applications in optoelectronics and data storage. With these developments approaching towards industrial applications, there will be a bright future for CNT-based nanoelectronics.

REFERENCES

1. R.S. Ruoff, D.C. Lorents, *Carbon* 33, 925–930 (1995).
2. P. Kim, L. Shi, A. Majumdar, P.L. McEuen, *Phys. Rev. Lett.* 87, 215502 (2001).
3. A. Javey, H. Kim, M. Brink, Q. Wang, A. Ural, J. Guo, P. Mcintyre, P. McEuen, M. Lundstrom, H. Dai, *Nat. Mater.* 1,241 (2002).
4. A. Javey, J. Guo, Q. Wang, M. Lundstrom, H. Dai, *Nature* 424, 654 (2003).
5. S.J. Tans, A.R.M. Verschueren, C. Dekker, *Nature* 393, 49 (1998).
6. M. Radosavljevic, M. Freitag, K.V. Thadani, A.T. Johnson, *Nano Lett.* 2, 761 (2002).
7. J. Kong, N. Franklin, C. Zhou, M. Chapline, S. Peng, K. Cho, H. Dai, *Science* 287, 622 (2000).
8. J.A. Robinson, E.S. Snow, S.C. Badescu, T.L. Reinecke, F.K. Perkins, *Nano Letters* 6, 1747–1751 (2006).
9. K.G. Ong, K. Zeng, C.A. Grimes, *IEEE Sens. J.* 2, 82 (2002).
10. Y. Lin, F. Lu, J. Wang, *Electroanalysis*, 16, 145 (2004); S. Sotiropoulou, N.A. Chaniotakis, *Anal. Bioanal. Chem.*, 375, 103 (2003); J. Wang, *Electroanalysis*, 17, 7 (2004).
11. D. Cai, J.M. Mataraza, Z.H. Qin, Z. Huang, J. Huang, T.C. Chiles, D. Carnahan, K. Kempa, Z. Ren, *Nat. Methods* 2, 449 (2005).
12. R. Singh, D. Pantarotto, D. McCarthy, O. Chaloin, J. Hoebeke, C.D. Partidos, J.P. Briand, M. Prato, A. Bianco, K. Kostarelos, *J. Am. Chem. Soc.* 127, 4388 (2005).
13. K.H. Park, M. Chhowalla, Z. Iqbal, F. Sesti, *J.Bio. Chem.* 278, 50212 (2003).
14. K.A. Shah, J.R. Dar, *Int J. Innov Res. Sci. Eng. Technol.* 3, 17395 (2014).
15. K.A. Shah, M.S. Parvaiz, *Superlattices Microstruct.* 93, 234 (2016).
16. Y.T. Yang, R.X. Ding, J.X. Song, *Phys. B* 406, 216 (2010).
17. S. Choudary, S. Qureshi, *Phys. Lett. A* 375, 3382 (2011).
18. .Taguchi, N. Igawa, H. Yamamoto, S. Jitsukawa, *J. Am. Ceram. Soc.* 88, 459 (2005).
19. J. Song, Y. Yang, H. Liu, *IEEE EDSSC* 60, 509 (2009).
20. K.A. Shah, M.S. Parvaiz, *Superlattices Microstruct.* 100, 375 (2016).
21. A. Loiseau, P. Launois, P. Petit, S. Roche, J.–P. Salvetat, Eds., *Understanding Carbon Nanotubes: From Basics to Applications*, Series Lecture Notes in Physics 677, ISBN: 3-540-26922-3, Springer (Berlin, 2006).
22. S. Reich, C. Thomsen, J. Maultzsch, *Carbon Nanotubes: Basic Concepts and Physical Properties*, ISBN3-527-40386-8, Wiley-VCH (Weinheim, 2004).
23. R. Saito, G. Dresselhaus, M.S. Dresselhaus, *Physical Properties of Carbon Nanotubes*, Imperial College Press (London, 1998).
24. K. Alam, R.K. Lake, *Appl. Phys. Lett.* 87, 073104 (2005).
25. K. Alam, R.K. Lake, *J. Appl. Phys.* 98, 064307 (2005).
26. J. Borghetti, V. Derycke, S. Lenfant, P. Chenevier, A. Filoramo, M. Goffman, D. Vuillaume, J.-P. Bourgoin, *Adv. Mater.* 18, 2535 (2006).
27. Quantum ATK version P-2019.03, Synopsis Quantum ATK (https://www.synopsys.com/silicon/quantumatk.html).
28. M.R. Pederson, J.Q. Broughton, *Phys. Rev. Lett.* 69, 2689 (1992).
29. D.S. Bethune, C.H. Kiang, M.S.D. Vries, G. Gorman, R. Savoy, J. Vazquez, R. Beyers, *Nature* 363, 605 (1993).
30. W. Smith, M. Monthioux, D.E. Luzzi, *Nature* 296, 323 (1998).
31. J. Sloan, J. Hammer, M.Z. Sibley, M.L.H. Green, *Chem. Commun.* 3, 347 (1998).
32. C.H. Kiang, J.S. Choi, T.T. Tran, A.D. Bacher, *J. Phys. Chem. B* 103, 7449 (1999).
33. P. Corio, A.P. Santos, M.L.A. Temperini, V.W. Brar, M.A. Pimenta, M.S. Dresselhaus, *Chem. Phys. Lett.* 383, 475 (2004).
34. A. Govindaraj, B.C. Satishkumar, M. Nath, C.N.R. Rao, *Chem. Mater.* 12, 205 (2000).

35. J. Sloan, D.M. Wright, H.G. Woo, S.R. Bailey, G. Brown, A.P.E. York, K.S. Coleman, J.L. Hutchison, M.L.H. Green, *Chem. Commun.* 700, 699 (1999).
36. R.S. Lee, H.J. Kim, J.E. Fischer, A. Thess, R.E. Smalley, *Nature* 388, 255 (1997).
37. S. Kazaoui, N. Minami, R. Jacquemin, H. Kataura, Y. Achiba, *Phys. Rev. B* 60, 13339 (1999).
38. E. Hern'andez, C. Goze, P. Bernier, A. Rubio, *Phys. Rev. Lett.* 80, 4502 (1999).
39. E. Hern'andez, C. Goze, P. Bernier, A. Rubio, *Appl. Phys. A. Mater.* 68, 287 (1999).
40. R.P. Gao, Z.L. Wang, Z.G. Bai, W.A. de Heer, L.M. Dai, M. Gao, *Phys. Rev. Lett.* 85, 622 (2000).
41. R.J. Baierle, S.B. Fagan, R. Mota, A.J.R. da Silva, A. Fazzio, *Phys. Rev.* 64, 085413 (2001).
42. E. Cruz-Silva, D. Cullen, L. Gu, J. M. Romo-Herrera, E. Mun˜oz-Sandoval, F. L´opez-Ur'ıas, D.J. Smith, H. Terrones, M. Terrones, *ACS Nano* 2, 441 (2008).
43. M.V. Kharlamova et al., *Eur. Phys. J. B* 84, 34 (2012).
44. Y. Chen, R.C. Haddon, S. Fang, A.M. Rao, P.C. Eklund, W.H. Lee, E.C. Dickey, E.A. Grulke, J.C. Pendergrass, A. Chavan, B.E. Haley, R.E. Smalley. *J. Mater. Res.* 13(9), 2423–2431, (1998).
45. M. Holzinger, O. Vostrowsky, A. Hirsch, F. Hennrich, M. Kappes, R. Weiss, F. Jellen, *Angew. Chem., Int. Ed.* 40(21), 4002–4005, (2001).
46. J. Chen, M.A. Hamon, H. Hu, Y. Chen, A.M. Rao, P.C. Eklund, R.C. Haddon. *Science* (Washington, DC), 282(5386), 95–98, (1998).
47. R. Weiss, S. Reichel, M. Handke, F. Hampel. *Angew. Chem. Int. Ed.* 37, 344, (1998).
48. S. Reichel. University of Erlangen-Nürnberg, PhD (1998).
49. R. Weiss, S. Reichel, *Eur. J. Inorg. Chem.* 1935, (2000).
50. P.J. Fagan, P.J. Krusic, C.N. McEwen, J. Lazar, D.H. Parker, N. Herron, E. Wasserman. *Science*, 262, 404, (1993).
51. M.S.P. Shaffer, K. Koziol. *Chem. Commun.* 2074–2075, (2002).
52. Y. Ying, R.K. Saini, F. Liang, A.K. Sadana, W.E. Bulleps. *Org. Lett.* 5, 1471–1473, (2003).
53. H. Xia, Q. Wang, G. Qiu, *Chem. Mater.* 15, 3879–3886, (2003).
54. F. Liang, A.K. Sadana, A. Peera, J. Chattopadhyay, Z. Gu, R.H. Hauge, W.E. Billups, *Nano Lett.* 4, 1257–1260, (2004).
55. J. Chattopadhyay, A.K. Sadana, F. Liang, J.M. Beach, Y. Xiao, R.H. Hauge, W.E. Billups. *Org. Lett.* 7, 4067–4069, (2005).
56. G. Viswanathan, N. Chakrapani, H. Yang, B. Wei, H. Chung, K. Cho, C. Y. Ryu, P.M. Ajayan, *J. Am. Chem. Soc.* 125, 9258–9259, (2003).
57. R. Blake, Y.K. Gun'ko, J.N. Coleman, M. Cadek, A. Fonseca, J.B. Nagy, W.J. Blau, *J. Am. Chem. Soc.* 126, 10226–10227, (2004).
58. S. Chen, W. Shen, G. Wu, D. Chen, M. Jiang. *Chem. Phys. Lett.* 402, 312–317, (2005).
59. M. Maggini, G. Scorrano, M. Prato, *J. Am. Chem. Soc.* 115, 9798–9799, (1993).
60. M. Prato, M. Maggini, *Acc. Chem. Res.* 31, 519–526, (1998).
61. S.E. Kooi, U. Schlecht, M. Burghard, K. Kern. *Angew. Chem. Int. Ed.* 41, 1353–1355, (2002).
62. F. Hauke, A. Hirsch, *J. Chem. Soc. Chem. Commun.* 21, 2199, (1999).
63. T. Nakajima, S. Kasamatsu, Y. Matsuo, *Eur. J. Solid State Inorg. Chem.* 33, 831 (1996).
64. E.T. Mickelson, C.B. Huffman, A.G. Rinzler, R.E. Smalley, R.H. Hauge, J.L. Margrave, *Chem. Phys. Lett.* 296, 188 (1998).
65. K.N. Kudin, H.F. Bettinger, G.E. Scuseria, *Phys. Rev. B Condens. Matter Mater. Phys.* 63, 045413 (2001).
66. P.J. Boul, J. Liu, E.T. Mickelson, C.B. Huffman, L.M. Ericson, I.W. Chiang, K.A. Smith, D.T. Colbert, R.H. Hauge, J.L. Margrave, R.E. Smalley, *Chem. Phys. Lett.* 310, 367–72 (1999).

67. K.F. Kelly, I.W. Chiang, E.T. Mickelson, R.H. Hauge, J.L. Margrave, X. Wang, G.E. Scuseria, C. Radloff, N.J. Halas, *Chem. Phys. Lett.* 313, 445–450 (1999).
68. R.K. Saini, I.W. Chiang, H. Peng, R.E. Smalley, W.E. Billups, R.H. Hauge, J.L. Margrave, *J. Am. Chem. Soc.* 125, 3617–3621 (2003).
69. S. Niyogi, M.A. Hamon, H. Hu, B. Zhao, P. Bhomik, R. Sen, M.E. Itkis, R.C. Haddon, *Acc. Chem. Res.* 35, 1105 (2002).
70. J.L. Bahr, J.M. Tour, *J. Mater. Chem.* 12, 1952–1958, (2002).
71. A. Hirsch. *Angew. Chem. Int. Ed.* 41(11), 1853–1859, (2002).
72. S. Banerjee, M.G.C. Kahn, S.S. Wong, *J. Chem. Eur.* 9, 1898–1908, (2003).
73. C.A. Dyke, J.M. Tour, *J. Phys. Chem. A* 108, 51, (2004).
74. S. Banerjee, T. Hemraj-Benny, S.S. Wong. *Adv. Mater.* 17(1), 17–29, (2005).
75. K. Balasubramanian, M. Burghard. *Small*, 1(2), 180–192, (2005).
76. R. Saito, G. Dresselhaus, M.S. Dresselhaus, *Physical Properties of Carbon Nanotubes*, Imperial College Press (London, 1998), pp. 17–33.
77. T. Hertel, G. Moos, *Phys. Rev. Lett.* 84, 5002 (2000).
78. J.-C. Charlier, X. Blase, S. Roche, *Rev. Mod. Phys.* 79, 677–732 (2007).
79. C. Zhou, J. Kong, E. Yenilmez, H. Dai, *Science* 290, 1552–1555 (2000).
80. G. Buchs, P. Ruffieux, P. Groning, O. Groning, *Appl. Phys. Lett.* 93, 073115 (2008).
81. B. Chandra, J. Bhattacharjee, M. Purewal, Y.-W. Son, Y. Wu, M. Huang, H. Yan, T.F. Heinz, P. Kim, J.B. Neaton, J. Hone, *Nano Lett.* 9, 1544–1548 (2009).
82. S.W. Lee, A. Kornblit, D. Lopez, S.V. Rotkin, A.A. Sirenko, H. Grebel, *Nano Lett.* 9, 1369–1373 (2009).
83. J. Li, Q. Zhang, *Carbon* 43, 667–670 (2005).
84. M. Ahlskog, O. Herranen, A. Johansson, J. Leppa¨niemi, D. Mtsuko, *Phys. Rev. B* 79, 155408 (2009).
85. Y. Xue et al., *Phys. Rev. B* 59, R7852 (1999).
86. N.D. Lang, *Phys. Rev. B* 55, 9364 (1997).
87. I.W. Lyo, Ph. Avouris, *Science* 245, 1369 (1989).
88. L. Chen et al., *Phys. Rev. Lett.* 99, 146803 (2007).
89. J. Chen, M.A. Reed, A.M. Rawlett, J.M. Tour, *Science* 286, 1550 (1999).
90. Y. Wu, D.B. Farmer, W. Zhu, S.-J. Han, C.D. Dimitrakopoulos, A.A. Bol, P. Avouris, Y.M. Lin, *ACS Nano* 6, 2610 (2012).
91. J. Zang, S. Ryu, N. Pugno, Q. Wang, Q. Tu, M.J. Buehler, *Nat. Mater.* 12, 1 (2013).
92. H. Tao, K. Yanagisawa, C. Zhang, T. Ueda, A. Onda, N. Li, T. Shou, S. Kamiya, J. Tao, *Cryst. Eng. Comm.* 14, 3027 (2012).
93. C. Mattevi, H. Kim, M. Chhowalla, *J. Mater. Chem.* 21, 3324–3334 (2011).
94. K.A. Shah, J.R. Dar, *Chin. J. Phys.* 55, 1142 (2017).
95. M. Radosavljevic, S. Heinze, J. Tersoff, P. Avouris, *Appl. Phys. Lett.* 83, 2435 (2003).
96. K.A. Shah, M.S. Parvaiz, G.N. Dar, *Phys. Lett. A* 383, 2207 (2019).
97. J. Appenzeller, J. Knoch, R. Martel, V. Derycke, S. Wind, P. Avouris, *Tech. Dig. Int. Electron Devices Meet* 11, 285 (2002).
98. P. Avouris, J. Appenzeller, R. Martel, S.J. Wind, *Proc. IEEE* 91, 1772 (2003).
99. P.L. McEuen, M.S. Fuhrer, H.K. Park, *IEEE Trans. Nanotechnol.* 1, 78 (2002).
100. M. Radosavljevic, J. Appenzeller, Ph. Avouris, J. Knoch, *Appl. Phys. Lett.* 84, 3693 (2004).
101. A. Javey, R. Tu, D.B. Farmer, J. Guo, R.G. Gordon, H. Dai, *Nano Lett.* 5, 345 (2005).
102. W. Gao, A. Kahn, *J. Phys. Condens. Matter.* 15, 2757 (2003).
103. L.M. Penga, Z. Zhang, S. Wang, X. Liang, *AIP Adv.* 2, 041403 (2012).
104. A. Kahn, N. Koch, W. Gao, *J. Polym. Sci. Part B Polym. Phys.* 41, 21 (2003).
105. G. Gruner, *Anal. Bioanal. Chem.* 384, 322 (2006).

4 Field-Effect Transistors Based on Graphene and Other Popular Two-Dimensional Materials

4.1 INTRODUCTION

Scaling of the transistors has prolonged the survival of Moore's law by more than a decade or so. However, the aggressive scaling of the devices has led to several short-channel effects, which eventually result in a degradation of the performance of the device in general. Furthermore, the heat dissipation involved in the devices has risen to new heights because of this scaling. These short-channel effects have prompted the scientific community to look for novel materials for device implementation. Out of these novel materials, two-dimensional (2D) materials and transition metal dichalcogenides (TMDs) have been widely investigated. Synthesis of graphene in 2004 [1] was an initiating step towards the exploration of the 2D materials and TMDs. Although the carrier mobility of graphene is very high [2], its application to device implementation is limited by the absence of a bandgap in it. TMDs are a class of materials consisting of a layer of transition material sandwiched between the layers of chalcogenide atoms. In general, TMDs are represented by the formula MX_2 where "M" represents the transition metal atom (Mo, W, Nb, Ta), and "X" represents the chalcogenide atom (S, Se, Te). TMDs are a family of materials with a wide variety of electronic and optical properties [3]. TMDs are the 2D semiconductors which have been intensively studied as potential replacement for silicon [4–7], with each of its variant material showing different electronic properties [8–12]. The layered structure of TMDs provides flexible control with high precision (atomic level) up to monolayer limit. The scaling in thickness is extremely desired for good electrostatic control of gate in ultra-short transistors. Monolayer MoS_2 has been theoretically shown to be better than silicon at sub 5 nm scaling limit [13,14]. Furthermore, single-layer TMDs from the past decade have attracted immense research interest because of their narrow band gap ranging from 1.1 to 2.0 eV for $MoSe_2$ and WS_2, respectively [15–17], direct to indirect transition of band gap [18–20], and hydrogen development while using as a catalyst. These materials thereby allow a wide variety of applications ranging from nanoelectronics to optoelectronics and energy conversion [21–26]. The amount of infrastructure invested in the silicon design is overwhelmingly large, and this becomes a reason for industries to be hesitant in attempting a material other than the silicon for nanoelectronics to optoelectronics applications. This has attracted the attention of researchers towards the 2D version of silicon, i.e., silicene. Silicene in its natural form has zero bandgap, and therefore, several methods have been given by the researchers for the generation of bandgap in silicene. Like silicene, germanene is the 2D version of germanium. In this chapter, field-effect transistors (FETs) and their applications based on these 2D materials reported in the literature have been revisited and are presented in the sections.

4.2 GRAPHENE FIELD-EFFECT TRANSISTORS (GFETs)

Graphene is a 2D sheet of single carbon atom wide with sp^2 hybridization [27]. Its excellent mechanical thermal and electrical properties have attracted extensive research interest [28]. Films consisting of graphene over flexible substrate are highly transparent and flexible [29–32]. The ultimate thinness in graphene, carbon–carbon bond, its planer nature, and band gap introduction by electron confinement in graphene nanoribbon; furthermore, very high electron mobility in graphene-based devices has resulted in great promise for application in nanoscale regime such as display devices, solar cells, energy storage, and sensors [33–41].

The complementary operation of conventional FETs helps in reduction of static power consumption. The off state in FETs arises due to the barrier potential between drain and source terminals. Also the current in the off state mode is proportional to ($e^{-Eg/kT}$). Here, E_g represents the band gap, since due to zero band gap in graphene and thus no barrier, which will result in unacceptable large current. Quantum confinement which means patterning of graphene into nanoribbons can be used to obtain band gap in graphene [42]. Graphene nanoribbons are found to show variation in band gap with the ribbon width by relation $E = \alpha/W$, α is found to be 2 eV, and W represents the ribbon width. Graphene nanoribbons with width around 5 nm yield substantial band gap to reduce significantly the thresh hold leakage. There are many other methods of band gap opening in graphene like substrate-induced band gap creation by crystal symmetry breakdown [43], bilayer graphene introduced to asymmetric field [44], and graphene obtained from nanomesh etching [45]. In non-FET switching mode, transistor no more requires bandgap for its operation. One of the approaches is to use the Klein tunneling for gapless graphene switch. Furthermore, bandgap-introduced graphene can also be used in tunnel FETs.

The armchair GNR shows increase in bandwidth with the decrease in width but increase in effective mass as well, which reduces the carrier velocity and hence reduces the carrier mobility. The variation of effective mass and band width with ribbon width is shown in Figure 4.1. In other materials, the electrical resistivity too shows size dependency in nanometer regime, e.g., InAs nanowire [46] and copper [47]. When the scaling of copper wire is done below 20 nm, Grain and edge boundary scattering leads to three times increase in resistivity in comparison with the bulk material. There are many reports of impact on graphene nanoribbon transport with edge scattering. Edge defects are responsible for breaking the perfect hexagonal repetition which defines a particular orientation [48]. The transport gap is found to be inversely proportional to nanoribbon width.

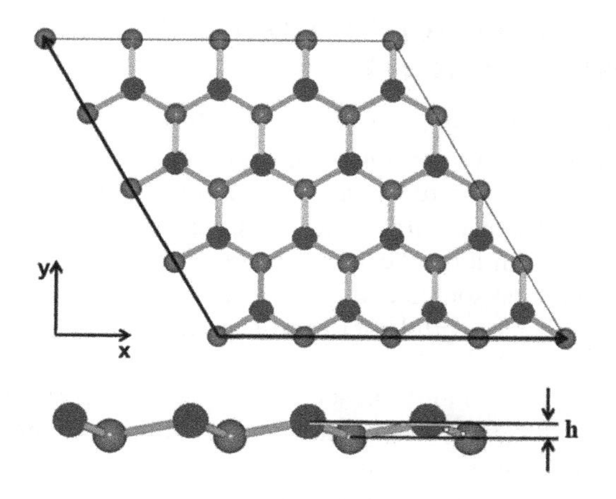

FIGURE 4.1 Buckled geometry of silicene (germanene).

With the increase in edge disorder, there is an increase in transport gap which is due to increase in lateral confinement. Even in metallic zigzag nanoribbons, the transport gap arise in small widths [49], which reveals that the transport gap in devices results in low mobility and does not depend on chirality, but for the best performance of the device, bandgap is preferred.

Length scaling is one of the important parameters as it is responsible for faster switching and in achieving high device density. In graphene, length scaling has indicated that the mobility remains the same down to $L = 2$ mm and shows decline when L is reduced below 2 mm [50]. This behavior was expected to be due to non-scaling in series resistance and the change to ballistic regime from diffusive regime of transport, and the extraction of mean free path to 300 ± 100 nm. It was reported that with the minimized interlayer coupling and very thin gate oxide FETs based on multilayer graphene achieved large on-current [51]. For the armchair graphene nanoribbon having a width of 2.7 nm, the band gap shows decline from 516 to 197 meV when the number of graphene layers were increased from one to ten layers. Furthermore, E_g shows monotonic decrease with the increase in coupling strength. The decrease in E_g is more prominent with large number of layers and also increases band-to-band tunneling, which is responsible for higher I_{OFF}.

Doping is one of the effective ways of conductivity control in graphene and also is responsible for the control of carrier type. It also helps to tone the carrier injection from the electrode to semiconducting material and alignment in bands between semiconductor and electrode. Furthermore, semiconductor is the prominent candidate for carrier injection. A variety of materials have been used for doping of graphene, but unlike substitutional doping as in case of conventional semiconductors, in graphene the doping is of surface charge transfer nature. It was found that the flaked graphene was lightly p-doped. The p-type doping was expected to be from the water vapor or oxygen in the surrounding. It was seen that CO and H_2O doping results in n-type and NO_2 and H_2O doping results in p-type with the concentration of gas in ppm range [52].

4.3 MOLYBDENUM DISULFIDE FIELD-EFFECT TRANSISTORS (MoS_2-FETs)

As the silicon-based transistors reach their scaling limit, due to short-channel effects [53], there arises an utmost need of exploring channel material for replacement of silicon [54–56]. The tunneling between the source to drain and the degradation of off-state leakage currents due to electrostatic gate control failure are responsible for scaling limit in silicon transistors [57,58]. The magnitude of the tunneling effect between the source and drain is reduced when the silicon channel is replaced by semiconductor with larger band-gap, heavier effective mass of carriers and smaller in plane dielectric constant [59]. For improved electrostatic control of gate, atomically narrow and uniform semiconductors with small in plane dielectric constant are required. Hence, introduction and exploration of semiconductors with enhanced properties than silicon could result in lower power dissipation in off state and more scaling of dimensions of transistor.

MoS_2 is being widely used as a catalyst in hydro sulfurization reaction and as solid lubricant [60–63] and as anode catalyst in oxide fuel cells with high temperature [64]. The MoS_2 forms a graphene like hexagonal structure of S and Mo atoms to form S-Mo-S sandwich. The S-Mo-S entities are arranged at the top of one other and are held with one another by weak non-covalent interactions. Every Mo atom is bonded covalently with six sulfur atoms, whereas every sulfur atom is attached with three Mo atoms. The bulk unit cell of MoS_2 corresponds to space group P63/mmc and consists of six (four sulfur and two Mo) atoms. The MoS_2 structure is presented by hexagonal lattice constant a, lattice constant c (out-of-plane), and internal displacement parameter z. The lattice constant and the displacement parameter are determined by (a = 3.16 Å), (c = 12.58 Å), and (z = 0.12). In the formation of MoS_2 sheet in bulk MoS_2, no covalent interactions plays vital role. In MoS_2, weak interlayer binding allows the separation of single/multilayer MoS_2 with the help of mechanical exfoliation and is the reliable method of obtaining MoS_2 [11,65].

Theoretical studies have revealed that there is transition from indirect band gap to direct band gap when the MoS_2 shrinks to single layer [66–68]. With the decrease of layers from bulk to single layer, there is an increase in band gap and the single-layer MoS_2 becomes direct bandgap semiconductor. It has been found that there is transition of bandgap from infrared region to visible region when moving from bulk material to monolayer MoS_2, which suggests its potential applications in optoelectronics, photonics, and sensing.

Desai et al. [69] has shown the reduction in off state source-to-drain tunneling current of the order of two related to silicon that is due to higher effective mass of electron in the transport direction, which in turn results in lower ballistic on current. They also demonstrated the transistor based on MoS_2 consisting of 1 nm physical gate length using carbon nanotube as gate electrode. The device shows outstanding switching characteristics and ideal threshold swing of around 65 mV/decade and the ratio of on current to off current of approximately 65. Also the effective channel length is found to vary from 3.9 to 1 nm for off state and on state, respectively.

The application of thin-layer MoS_2 transistor has been demonstrated in optoelectronic devices [70,71] and logic devices [7,72]. The MoS_2 transistor has performed exceptionally well at high frequencies as well. The top-gated MoS_2 transistor has been demonstrated operating in the frequency range of gigahertz [73], which shows cutoff frequency of almost 6GHz. Furthermore, the existence of band gap leads to current saturation [74], which allows power and voltage gains. These advantages prove that the device based on MoS_2 material will prove a promising candidate for high-speed amplifier, logic realization, etc.

4.4 MOLYBDENUM DISELENIDE FIELD-EFFECT TRANSISTORS ($MoSe_2$-FETs)

The interest towards monolayer transition metal chalcogenides has renewed after success in graphene, and it rejuvenated after the pioneer paper in 2011 [21] in which the top-gated MoS_2 FET at room temperature was demonstrated, where exfoliated layer of $MoSe_2$ of thickness 6.5 Å was placed as semiconducting channel was deposited over SiO_2 and covered with HfO_2 of thickness 300nm which functioned as top-gated dielectric. A high mobility of 200 cm^2/V s was achieved. Single-layer $MoSe_2$ presents very good thermal stability and direct band gap of 1.55eV as reported from photoluminescence measurements. The photoluminescence intensity in peak increases intensely from many layers to monolayer which is the result of transition from indirect bandgap to direct bandgap similar to MoS_2 behavior [18,19,75]. More importantly, the MoS_2 flakes have almost degenerate indirect and direct bandgap and the rise in temperature can force the system to the quasi-2D limit by reducing thermally the coupling in between layers. $MoSe_2$ shows slight difference when compared with MoS_2; furthermore, the bandgap in $MoSe_2$ is fine matched with solar spectrum. Hence, it promotes new direction for use in 2D applications in which external modification of bandgap and optical properties are desired.

$MoSe_2$ in its bulk form is indirect semiconductor with a bandgap of 1.1 eV and increases to 1.55eV but direct band gap semiconductor in single layer and bilayer [76,77]. The $MoSe_2$-based FETs are n-type and hold well gate control and the ratio of on to off current greater than 10^6. The dependency of temperature tells that with the increase in temperature, the mobility decreases which justifies phonon scattering at room temperature [78]. Chamlagain et al. [79] reported scanning tunneling microscopy characterization in $MoSe_2$ crystal at low temperature. They fabricated $MoSe_2$ FET on parylene-C and SiO_2 substrate and calculated the electrical characteristics. The mobility in $MoSe_2$-based device over parylene-C has been found to be close to mobility in bulk $MoSe_2$ ~100–160 cm^2/V s, which is higher than that of MoS_2-based device over SiO_2 substrate ~50 cm^2/V s. The mobility at room temperature shows thickness independency in both substrates. The metal–insulator transition has been predicted with characteristic conductivity of e^2/h by transport measurement at varied temperature. The mobility in $MoSe_2$ is found to increase up to ~500 cm^2/V s for both parylene-C and SiO_2 (substrate based) as temperature decreases to ~100K; furthermore, the mobility of $MoSe_2$ has been found to increase more rapidly. The mobility of $MoSe_2$ device over SiO_2 substrate at temperature greater than 200K is almost independent of charge impurity. The mobility dependency on substrate is attributed to surface roughness scattering.

Pradhan, Rhodes, Xin et al. [80] reported a study on few-layer $MoSe_2$-based FET fabricated by chemical vapor transport technique exfoliated mechanically over SiO_2. $MoSe_2$ FET with Ti electrical contact shows ambipolar behavior and the ratio of on current to off current of around 10^6 for both the electron and hole channels when subjected to tiny excitation voltage. For both electron and hole channels, the Hall effect designates the Hall mobility μH ~250 cm^2/(V s), which is as good as field-effect mobilities (μ = 150–200 cm^2/(V s)) assessed through the two-terminal field-effect configuration. Hence, from the above mentioned results, the $MoSe_2$ could prove a better candidate for p–n junction for low-power complementary digital logic with single atomic layer.

4.5 TUNGSTEN DISULFIDE FIELD-EFFECT TRANSISTORS (WS_2-FETs)

Tungsten disulfide (WS_2) is the other member of TMDs, which is a semiconductor. In its monolayer form, WS_2 is a direct bandgap semiconductor having a bandgap of 2.0 eV [81]. Based on the theoretical results, WS_2 is predicted to have the highest mobility of the carriers among the semiconducting members of the TMD family [82]. From the material point of view, some of the advantages of WS_2 include high thermal stability and absence of dangling bonds [83]. Monolayer WS_2 is fast-finding its domain in applications such as optoelectronic devices, memory devices, and chemical sensors. The mobility of carriers in monolayer WS_2 at room temperature has been reported to range from 40 to 60 cm^2/V s [84,85]. The transport properties of monolayer WS_2 were reported to be affected by surface roughness and impurities on Si/SiO_2 substrates [86,87]. Using a suspended geometry may result in the reduction of these problems [88] at the cost of fabrication constraints. Because of the exciting properties of TMDs in general and WS_2 in particular, a number of devices based on WS_2 are reported in the literature. A brief discussion of the devices based on WS_2 is given next.

Wan Sik Hwang et al. have reported the realization of a FET based on WS_2 in multilayer form [83]. An appreciable current on/off ratio of 10^5 is achieved in this device at a drain-to-source voltage of 1 V. Once the drain-to-source voltage is increased to 5 V, the on/off current ratio comes down to 10^4. The gate leakage current of the device is reported to be in the range of a few Pico amperes, which is very small as compared to the drain current. An accumulation of electrons (holes) for positive gate voltages (negative gate voltages) is observed. The observed behavior is a clear proof of the ambipolarity in the device. Such a behavior of ambipolarity can be advantageous for the inverter applications of the CMOS technology [89]. Besides, the work in Ref. [83] also includes an investigation of the optoelectronic characteristics of WS_2 flakes and proposes the future use of WS_2 for optoelectronic applications.

Roi Levi et al. have reported the first transistor based on the WS_2 in tubular form [90]. In this work, nanotubes based on WS_2 have been fabricated, and later, a FET is fabricated from the said geometry of the WS_2. The mobility of the carriers in the device is reported to be 50 cm^2/V s. Furthermore, it was demonstrated that the current-carrying capacity of the WS_2 nanotubes can attain a value as high as 2.4×10^8 A/cm^2. The nanotubes used in the fabrication of the device had a diameter ranging from 50 to 200 nm, whereas the length of the nanotube ranges between 500 and 3,500 nm. The value of the differential conductivity in the reported device was estimated to be as high as $3\times10^3/\Omega$ cm. This value of the conductivity is only an underestimation of the exact value because the reported device never reached a current saturation because of the limitations of the experimental setup [90].

Hafiz et al. have reported a FET based on WS_2 using a chemical doping technique to achieve lower contact resistance [91]. After doping, the contact resistance showed a value of 0.9 kΩ·µm. This reduced value of the contact resistance results in a reduced Schottky barrier, which ultimately degrades the performance of the device. The reported device is characterized by the values of 1.05×10^6, 34.7 $cm^2/(V{\cdot}s)$, and 65 µA/µm for on/off current ratio, mobility of carriers, and on state current, respectively. Furthermore, the device is doped in an n-type fashion with the help of lithium fluoride (LiF). Besides improving the electrical performance of the device, doping with LiF also improves the photoelectrical properties of the FETs based on WS_2. The employed doping technique results in a relatively higher value of the photocurrent as compared to the un-doped version of the device based on WS_2.

Besides the applications to FETs, WS_2 has been actively used in the development of devices sensitive to light. Sanghyun Jo et al. have demonstrated a light-emitting transistor based on monolayer and bilayer WS_2 [92]. The devices demonstrated an ambipolar behavior arising due to the simultaneous injection of the electrons and holes from the two electrodes wherein the emission of the light from the FET channel was observed. The device is based on the ambipolar ionic liquid gated architecture. The reported work portrays the power of the ionic liquid gating technique. Such a technique can be used to evaluate the bandgap of the material using simple measurement of transport of the carriers and is compatible with the optical measurements. The technique can therefore be used for the ambipolar injection of the carriers in the device. The results reported in this work pave the way for applying this technique for evaluating the optoelectronic properties of the atomically thin materials like TMDs at a much broader range of the carrier density than what has been achieved before.

4.6 SILICENE AND GERMANENE FIELD-EFFECT TRANSISTORS

Progressive growth, understanding, and characterization of the electronic and physical properties of graphene has prompted the quest for other two-dimensional (2D) materials such as silicene, germanene, phosphorene, hexagonal boron nitride (hBN), and some heterostructures based on these materials. A lot of attention has been observed to silicene and germanene over the past decade or so because of their expected integration and compatibility to the present state of the art silicon nanotechnology. Extensive theoretical study has been focused on these materials, and recently, formation of silicene on metallic substrates such as Ag (111), [93–95] ZrB_2 (0001) [96], and Ir (111) [97] has been reported in the literature. Besides, the germanene formation on Au (111) [98] has also been reported.

Silicene (germanene) in its freestanding form comprises a single layer of Si (Ge) atoms arranged in a hexagonal fashion. Unlike graphene, silicene (germanene) has a buckled geometry wherein the two sub-lattices of the hexagon are separated by vertical distance (called buckling height h) of about 0.44 Å (0.66 Å) as shown in

Figure 4.1 [99]. Large atomic radius and large bond length in silicene (germanene) as compared to graphene prevent the Si (Ge) atoms to be purely sp^2-hybridized. Thus, the hybridization in silicene (germanene) is of mixed (sp^2 and sp^3) nature. This mixed hybridization eventually leads to a buckled geometry in silicene (germanene) [100]. Though similar in many respects, silicene and germanene have some structural differences, which are given in Table 4.1.

Like graphene, the buckled form of silicene (germanene) is a gapless semiconductor having a Dirac cone at the K-point in the Brillouin zone as shown in Figure 4.2 [99]. This buckled geometry of silicene (germanene) allows the modification of the electronic properties. For example, a perpendicular electric field to the plane of silicene (germanene) results in an energy gap at the K-point [101–104]. The bandgap is opened due to the charge transfer between the upper and lower Si (Ge) atoms, wherein the inversion symmetry of the system is broken under the influence of the perpendicular electric field. This field induced band gap in silicene (germanene) is of interest for logic applications. Besides the perpendicular electric field, other techniques like chemical functionalization with various ad-atoms [105–110] and mechanical strain [111,112] are also reported to result in the modification of the electronic properties of silicene (germanene). Because of the sp^3 hybridization, various atoms are easily adsorbed on silicene (germanene), hence leading to a bandgap in functionalized silicene (germanene). Furthermore, application of a biaxial strain (more than 5%) leads to hole doping in silicene (germanene) because of the reduced strength of Si–Si (Ge–Ge) bonds. Besides, a symmetry breaking mechanism is responsible for bandgap opening under the application of a uniaxial strain [112].

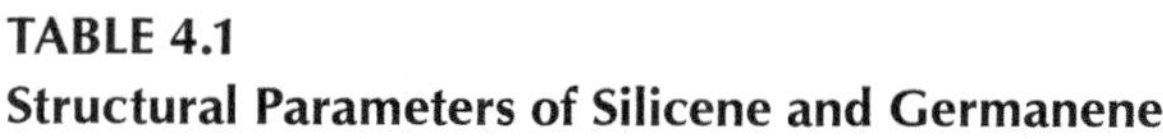

TABLE 4.1
Structural Parameters of Silicene and Germanene

	Lattice Parameter (Å)	Buckling Distance h (Å)	Bond Length (Å)
Silicene	3.87	0.44	2.28
Germanene	4.06	0.69	2.44

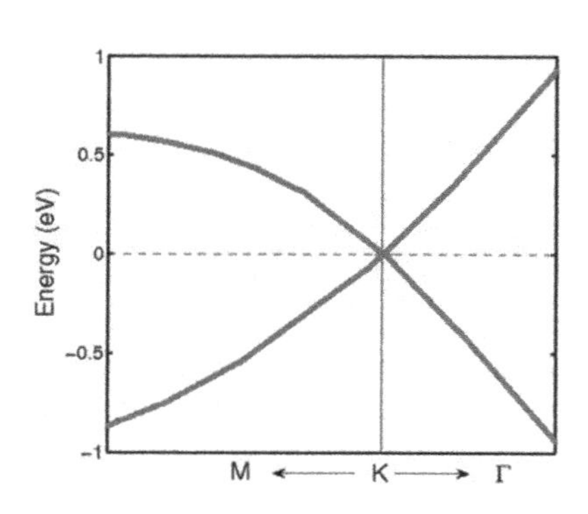

FIGURE 4.2 Dirac zone in silicene (LDA).

Other methods of modification of the electronic properties of silicene (germanene) include its growth on substrates such as Ag (111) [101,113] and Au (111) [98], wherein a semiconducting nature of silicene is observed with a bandgap of 0.3 eV.

The electronic properties of silicene (germanene) are quite promising, but the absence of a bandgap in it is the biggest hurdle in its development for device (FET) applications. The presence of a bandgap is a must in order to use a particular material for the application of a FET and allows for the switching of the device from its on state to off state. A number of methods including those reported above have a primary aim of introducing a bandgap in silicene (germanene) such that the exotic properties of the material may be used for development of more efficient devices and circuits. Opening of a sizeable bandgap without degrading the other electronic properties is of critical importance for the development of silicene based FETs. Theoretically, a number of FETs based on silicene are reported in the literature. These FETs in general adopt one of the methods reported above for bandgap generation. Here, we describe some of these FETs reported in the literature:

4.6.1 Dual-Gated Silicene FET

A FET based on silicene using a dual gate configuration and SiO_2 as dielectric was investigated theoretically by Liu's group [101]. The device uses a vertical electric field in order to control the switching (on to off and vice versa) mechanism of the device. The proposed device shows an on/off ratio of 4.2 at the room temperature under the application of a perpendicular electric field of magnitude $E_{\perp} = 1\,\text{V/Å}$. This appreciably low value of the on/off ratio is a direct consequence of the short length of the channel (~67 Å) chosen for the proposed device. Such a short-channel length results in a large leakage current in the off state and hence degrades the on/off ratio of the device. The reactive nature of silicene due to its sp^3 hybridization may lead to a covalent bonding with the Si and O atoms of the SiO_2 dielectric. Thus, a material which could act as a buffer between the silicene and SiO_2 is desirable for the proper operation of the silicene-based FET. hBN has been used as such a buffer material and has been proven that such a material remains intact even under the application of an electric field of magnitude 2 V/Å. Hence, insertion of a hBN layer is strongly recommended between the silicene and SiO_2 for silicene-based FETs.

4.6.2 Alkali-Adsorbed Silicene-Based FET

As stated above, adsorption of atoms results in a bandgap in silicene. If the atoms used are alkali metal atoms, the induced bandgap can attain a value of up to 0.5 eV in silicene. A FET based on Na atom adsorbed silicene channel is reported in [107]. The device comprises semi-infinite electrodes made of silicene, a dielectric region made of hBN and SiO_2 substrate, and a bottom gate. The device shows the switching characteristics at gate voltages of 0 V (on state) and –30 V (off state). At zero gate voltage V_g, the fermi level E_f is located above the transport gap. As V_g decreases to a negative value, E_f moves downwards, closer to the transport gap. At a V_g of –30 V, E_f lies in the transport gap and the transmission nearly vanishes within the transmission window consequently, resulting in an off state. In the on state, the transmission eigenvalue attains a value of 0.70, and the corresponding incoming wave traverses to the other electrode with very little scattering. On the contrary, the incoming wave in the off state does not reach the other electrode because of excessive scattering; furthermore, the transmission eigenvalue in the off state is only 0.01. The current on/off ratio of this device is as high as 4×10^8.

4.6.3 Silicene Nanomesh FET

Although a large bandgap (0.5 eV) can be attained by adsorption of alkali metal atoms, the FET based on the alkali metal adsorbed silicene requires a large supply voltage of around 30 V. Hence, a silicene FET with high on/off ratio and a reduced value of the supply voltage is desirable. Silicene nanomesh (SNM), another modified version of silicene comprising of a periodic array of hexagonal or triangular holes, is also a promising material for FET implementation [113]. In the SNM, the dangling bonds near the holes are passivated by hydrogen atoms. A maximum bandgap of 0.68 eV can be attained in SNMs. The FET based on SNM has some unique performance parameters with a current on/off ratio of 5.1×10^4 and a sub-threshold swing ($SS = dV_{gate}/d(\log I)$) of 68 mV/dec [114]. The reported work also evaluates the performance of the device with decreasing channel length. It is reported that on/off current ratio decreases from 5.1×10^4 at 9.1 nm to 17 at 3.8 nm. In addition, the value of SS increases from 68 mV/dec at 0.1 μm to 336 mV/dec at 3.8 nm. The degraded performance of the SNM FET at reduced channel lengths is a direct consequence of the short-channel effects. In general, the overall performance of the SNM FET at room temperature is excellent.

4.6.4 Silicene Nanoribbon FET

FET based on armchair silicene nanoribbon (ASiNR) is reported in [115]. The tunability of the properties of the FET by changing the width and length of the nanoribbon is presented in this work. It has been long proposed that the bandgap of the nanoribbons in general changes with the width. The bandgap of the silicene nanoribbons and thus the on/off ratio show a similar behavior. The highest value of the on/off current ratio in the evaluated devices can reach a value as high as 10^6. The lowest value of SS attained is 90 mV/dec; further, the subthreshold swing shows a monotonous decrease with the increase in the length of the nanoribbon. Since the device is based on a nanoribbon, there are two disadvantages: first, the nanoribbons are difficult to fabricate, and second, the silicene in its pristine form is unstable. These disadvantages can prove to be the biggest hurdle in the experimental development of the silicene nanoribbon-based FETs.

4.6.5 Li-Cl Co-Decorated Sub-10-nm Silicene Nanoribbon FET

An approach of co-decorating the zigzag-type SiNRs (ZSiNRs) with lithium and chlorine has been adopted for creating a bandgap in SiNRs in Ref. [116]. Among various combinations considered for generating a bandgap in SiNRs such as hydrogen–fluorine and oxygen–hydrogen, the authors have shown that the combination of

Li and Cl shows the best results as far as the sizable bandgap and the formation energy are considered. With Li–Cl co-decoration, it is found that the formation energy is of the order of −1.63 eV/Å having an optimum bandgap of 0.72 eV at a width of six atoms. Thus, Li–Cl co-decorated SiNR is reported to be a stable and relatively better choice material for FET implementation. The structure of Li–Cl decorated silicene FET is given in Figure 4.3.

4.6.6 Silicene Tunnel FET

The fundamental issue for nanoelectronics has been the power dissipation. Tunnel FETs (TFETs) based on band-to-band tunneling can have a subthreshold swing lesser than the classical limit of 60 mV/dec, operating at reduced values of the supply voltage as compared to the conventional FETs (MOSFETs). These reduced values of SS and supply voltage can allow for reduced power dissipation. Simulation of a TFET based on silicene, comprising of differently doped silicene in a P-I-N fashion is presented in [101]. The on state current attains a value of 1,000 μA/μm. Furthermore, the SS reaches a value of 77 mV/dec at a supply voltage of 1.7 V. In general, the TFETs based on conventional materials (like silicon) are characterized by feeble on currents. In order to enhance the on current, the supply voltage has to be greatly increased, thus, vindicating the overall idea of reducing the power dissipation by using TFETs. However, the theoretical values of the TFET stated above are quite promising in achieving the aim of reduced power dissipation.

4.7 SUMMARY

Two-dimensional (2D) materials are considered important for future CMOS technology. As discussed in this chapter, several 2D materials ranging from TMDs to silicene to graphene are under regressive research attention in order to exploit their best features for different nano-electronics and optoelectronics application. The main building of these applications is the FET. In this chapter, an effort was carried

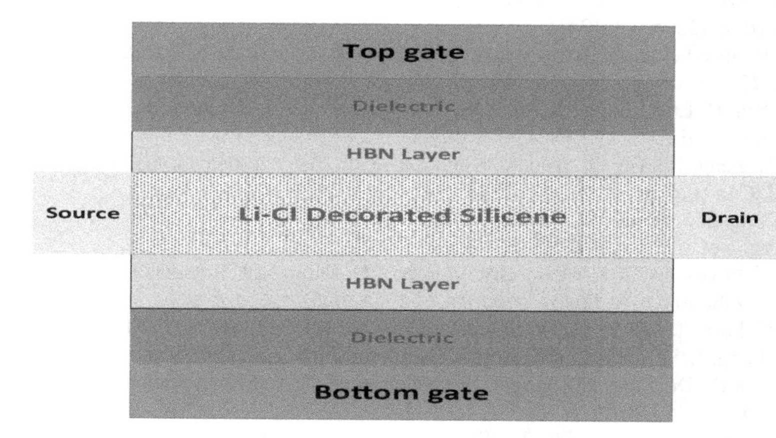

FIGURE 4.3 Model of the Li-Cl-decorated double-gate SiNR FET.

out to highlight the different FET structures reported through these 2D materials. A light was thrown on the fabrication progress, special features, challenges, and different applications.

REFERENCES

1. K.S. Novoselov, A.K. Geim, S.V. Morozov, D. Jiang, Y. Zhang, S.V. Dubonos, I.V. Grigorieva, A.A. Firsov, *Science* 306, 666–669 (2004).
2. K.I. Bolotin, K.J. Sikes, Z. Jiang, M. Klima, G. Fudenberg, J. Hone, P. Kim, H.L. Stormer, *Solid State Commun.* 146, 351–355 (2008).
3. Q.H. Wang, K. Kalantar-Zadeh, A. Kis, J.N. Coleman, M.S. Strano, *Nat. Nanotechnol.* 7, 699–712 (2012).
4. B. Radisavljevic, A. Radenovic, J. Brivio, V. Giacometti, A. Kis, *Nat. Nanotechnol.* 6, 150 (2011).
5. D. Sarkar et al., *Nature* 526, 91 (2015).
6. H. Liu, A.T. Neal, P.D. Ye, *ACS Nano* 6, 8563 (2012).
7. H. Wang et al., *Nano Lett.* 12, 4674 (2012).
8. K.F. Mak, K.L. McGill, J. Park, P.L. McEuen, *Science* 344, 1489 (2014).
9. D. Jariwala, V.K. Sangwan, L.J. Lauhon, T.J. Marks, M.C. Hersam, *ACS Nano* 8, 1102 (2014).
10. H. Fang et al., *Proc. Natl. Acad. Sci. U.S.A.* 111, 6198 (2014).
11. K.S. Novoselov et al., *Proc. Natl. Acad. Sci. U.S.A.* 102, 10451 (2005).
12. C.H. Lee et al., *Nat. Nanotechnol.* 9, 676 (2014).
13. Y. Yoon, K. Ganapathi, S. Salahuddin, *Nano Lett.* 11, 3768 (2011).
14. L. Liu, Y. Lu, J. Guo, *IEEE Trans. Electron. Dev.* 60, 4133 (2013).
15. A. Kuc, N. Zibouche, T. Heine, *Phys. Rev. B* 83, 245213 (2011).
16. Q.H. Wang, K. Kalantar-Zadeh, A. Kis, J.N. Coleman, M.S. Strano, *Nat. Nanotechnol.* 7, 699 (2012).
17. M. Chhowalla, H.S. Shin, G. Eda, L.J. Li, K.P. Loh, H. Zhang, *Nat. Chem.* 5, 263 (2013).
18. K.F. Mak, C. Lee, J. Hone, J. Shan, T.F. Heinz, *Phys. Rev. Lett.* 105, 136805 (2010).
19. A. Splendiani, L. Sun, Y. Zhang, T. Li, J. Kim, C-Y. Chim, G. Galli, F. Wang, *Nano Lett.* 10, 1271 (2010).
20. Z.Z. Wu, B.Z. Fang, A. Bonakdarpour, A.K. Sun, D.P. Wilkinson, D.Z. Wang, *Appl. Catal. B* 125, 59 (2012).
21. B. Radisavljevic, A. Radenovic, J. Brivio, V. Giacometti, A. Kis, *Nat. Nanotechnol.* 6, 147 (2011).
22. Z. Yin, H. Li, L. Jiang, Y. Shi, Y. Sun, G. Lu, Q. Zhang, X. Chen, H. Zhang, *ACS Nano* 6, 8563 (2012).
23. H.L. Zeng, J.F. Dai, W. Yao, D. Xiao, X.D. Cui, *Nat. Nanotechnol.* 7, 490 (2012).
24. L.J. Cao et al. *Small* 9, 2905 (2013).
25. G. Eda, S.A. Maier, *ACS Nano* 7, 5660 (2013).
26. J. Yoon et al. *Small* 9, 3295 (2013).
27. K.S. Novoselov, A.K. Geim, S.V. Morozov, D. Jiang, M.I. Katsnelson, I.V. Grigorieva, S.V. Dubonos, A.A. Firsov, *Nature* 438, 197 (2005).
28. A.C. Ferrari, J.C. Meyer, V. Scardaci, C. Casiraghi, M. Lazzeri, F. Mauri, S. Piscanec, D. Jiang, K.S. Novoselov, S. Roth, A.K. Geim, *Phys. Rev. Lett.* 97, 187401 (2006).
29. L. Gomez De Arco, Y. Zhang, C.W. Schlenker, K. Ryu, M.E. Thompson, C. Zhou, *ACS Nano* 4, 2865 (2010).
30. Y. Wang, R. Yang, Z. Shi, L. Zhang, D. Shi, E. Wang, G. Zhang, *ACS Nano* 5, 3645 (2011).

31. S. Bhaviripudi, X. Jia, M.S. Dresselhaus, J. Kong, *J. Nano Lett.* 10, 4128 (2010).
32. D. Sui, Y. Huang, L. Huang, J. Liang, Y. Ma, Y. Chen, *Small* 7, 3186 (2011).
33. D. Wei, Y. Liu, Y. Wang, H. Zhang, L. Huang, Y. Yu, *Nano Lett.* 9, 1752 (2009).
34. H.J. Jin, S.H. Lee, T.H. Kim, J. Park, H.Y. Song, T.H. Park, S. Hong, *Biosens. Bioelectron.* 35, 335 (2012).
35. H. Park, P.R. Brown, V. Bulovic, J. Kong, *Nano Lett.* 12, 133 (2012).
36. X. Wang, L. Zhi, K. Mullen, *Nano Lett.* 8, 323 (2008).
37. X. Liu, R. Aizen, R. Freeman, O. Yehezkeli, I. Willner, *ACS Nano* 6, 3553 (2012).
38. G. Yu, L. Hu, M. Vosgueritchian, H. Wang, X. Xie, J.R. McDonough, X. Cui, Y. Cui, Z. Bao, *Nano Lett.* 11, 2905 (2011).
39. H. Wang, Y. Yang, Y. Liang, J.T. Robinson, Y. Li, A. Jackson, Y. Cui, H. Dai, *Nano Lett.* 11, 2644 (2011).
40. Y.T. Liang, B.K. Vijayan, K.A. Gray, M.C. Hersam, *Nano Lett.* 11, 2865 (2011).
41. R.X. He, P. Lin, Z.K. Liu, H.W. Zhu, X.Z. Zhao, H.L.W. Chan, F. Yan, *Nano Lett.* 12, 1404 (2012).
42. J.D. Meindl, J.A. Davis, *IEEE J. Solid-State Circuits* 35, 1515 (2000).
43. V.V. Zhirnov, R.K. Cavin, J.A. Hutchby, G.I. Bourianoff, *Proc. IEEE* 91, 1934 (2003).
44. Y. Taur, *IBM J. Res. Dev.* 46, 213 (2002).
45. K.L. Wang, K. Galatsis, R. Ostroumov, M. Ozkan, K. Likharev, Y. Botros, "Nanoarchitectonics: Advances in nanoelectronics", in *Handbook of Nanoscience, Engineering and Technology*, 2nd ed, W.A. Goddard, III, W.D. Brenner, S.E. Lyshevski, J.G. Iafrate Eds. CRC Press (Boca Raton, FL, 2007).
46. M. Butts, A. DeHon, S.C. Goldstein, *IEEE/ACM International Conference on Computer Aided Design*. IEEE/ACM Digest of Technical Papers (Cat. No.02CH37391). IEEE Piscataway, NJ, USA, p. 433, 2002.
47. X. Wang, Y. Ouyang, X. Li, H. Wang, J. Guo, H. Dai, *Phys. Rev. Lett.* 100, 206803, (2008).
48. K.K. Likharev, D.B. Strukov, *Introducing Molecular Electronics*. Springer-Verlag (Berlin, 2005). p. 447.
49. J.E. Green, J. Wook Choi, A. Boukai, Y. Bunimovich, E. Johnston-Halperin, E. DeIonno, Y. Luo, B.A. Sheriff, K. Xu, Y. Shik Shin, H.-R. Tseng, J.F. Stoddart, J.R. Heath, *Nature* 445, 414 (2007).
50. T. Jayasekera, B.D. Kong, K.W. Kim, M. Buongiorno Nardelli, *Phys. Rev. Lett.* 104, 146801 (2010).
51. V. Karpan, G. Giovannetti, P. Khomyakov, M. Talanana, A. Starikov, M. Zwierzycki, J. van den Brink, G. Brocks, P. Kelly, *Phys. Rev. Lett.* 99, 176602 (2007).
52. I. Spielman, J. Eisenstein, L. Pfeiffer, K. West, *Phys. Rev. Lett.* 84, 5808 (2000).
53. M. Lundstrom, *Science* 299, 210 (2003).
54. T.N. Theis, P.M. Solomon, *Science* 327, 1600 (2010).
55. R. Chau, B. Doyle, S. Datta, J. Kavalieros, K. Zhang, *Nat. Mater.* 6, 810 (2007).
56. A. D. Franklin, *Science* 349, 2750 (2015).
57. M. Luisier, M. Lundstrom, D.A. Antoniadis, J. Bokor, *IEEE Int. Electron Devices Meet.* 11 (2011).
58. H. Kawaura, T. Sakamoto, T. Baba, *Appl. Phys. Lett.* 76, 3810 (2000).
59. W.S. Cho, K. Roy, *IEEE Electron Device Lett.* 36, 427 (2015).
60. P. Raybaud, J. Hafner, G. Kresse, S. Kasztelan, H. Toulhoat, *J. Catal.* 189, 129 (2000).
61. H. Schweiger, P. Raybaud, G. Kresse, H. Toulhoat, *J. Catal.* 207, 76 (2002).
62. L.S. Byskov, J.K. Norskov, B.S. Clausen, H. Topsoe, *J. Catal.* 187, 109 (1999).
63. M. Sun, J. Adjaye, A.E. Nelson, *Appl. Catal. A* 263, 131 (2004).
64. N.M. Galea, E.S. Kadantsev, T. Ziegler, *J. Phys, Chem. C* 113, 193 (2009).
65. R.F. Frindt, *J. Appl. Phys.* 37, 1928 (1966).
66. E. Scalise, M. Houssa, G. Pourtois, V. Afanas'ev, A. Stesmans, *Nano Res.* 5, 43 (2012).

67. H. Zeng, J. Dai, W. Yao, D. Xiao, X. Cui, *Nature Nanotech.* 7, 490 (2012).
68. T. Cao, G. Wang, W. Han, H. Ye, C. Zhu, J. Shi, Q. Niu, P. Tan, E. Wang, B. Liu, J. Feng, *Nature Commun.* 3, 887 (2012).
69. B.D. Sujay et al., *Device Technol.* 354, 6308 (2016).
70. M. Schroter, M. Claus, P. Sakalas, M. Haferlach, W. Dawei, *IEEE J. Electron Devices Soc.* 1, 9 (2013).
71. Y.M. Lin, K.A. Jenkins, A. Valdes-Garcia, J.P. Small, D.B. Farmer, P. Avouris, *Nano Lett.* 9, 422 (2008).
72. B. Radisavljevic, M. B. Whitwick, A. Kis, *ACS Nano* 5, 9934 (2011).
73. K. Daria et al., *Nano Lett.* 14, 5905–5911 (2014).
74. D. Lembke, A. Kis, *ACS Nano* 6, 10070 (2012).
75. G. Eda, H. Yamaguchi, D. Voiry, T. Fujita, M. Chen, M. Chhowalla, *Nano Lett.* 11, 5111 (2011).
76. S. Tongay, J. Zhou, C. Ataca, K. Lo, T.S. Matthews, J. Li, J.C. Grossman, J. Wu, *Nano Lett.* (published online).
77. W.S. Yun, S. Han, S.C. Hong, I.G. Kim, J.D. Lee, *Phys. Rev. B* 85, 033305 (2012).
78. L. Stefano, F. Babak, T Emanuel, *Appl. Phys. Lett.* 101, 223104 (2012).
79. Chamlagain, Li, Qing et al., *Am. Phys. Soc.*, APS March Meeting 2013, March 18–22, abstract id. T23.005, (2013).
80. N.R. Pradhan, D. Rhodes, Y. Xin, et al., *ACS Nano* 8, 7923–7929 (2014).
81. A. L. Elías et al., *ACS Nano* 7, 5235–5242 (2013).
82. L. Liu, S. B. Kumar, Y. Ouyang, J. Guo, *IEEE Trans. Electron Devices* 58, 3042–3047 (2011).
83. W.S. Hwang et al. *Appl. Phys. Lett.* 101, 013107 (2012).
84. S. Jo, N. Ubrig, H. Berger, A.B. Kuzmenko, A.F. Morpurgo, *Nano Lett.* 14, 2019–2025 (2014).
85. D. Ovchinnikov, A. Allain, Y.S. Huang, D. Dumcenco, A. Kis, *ACS Nano*, 8, 8174–8181 (2014).
86. F. Withers, T.H. Bointon, D.C. Hudson, M.F. Craciun, S. Russo, *Sci. Rep.* 4, 4967(2014).
87. D.J. Late, B. Liu, H.R. Matte, V.P. Dravid, C. Rao, *ACS Nano* 6, 5635–5641 (2012).
88. K.I. Bolotin et al., *Solid State Commun.*, 146, 351–355 (2008).
89. W.J. Yu, U.J. Kim, B.R. Kang, I.H. Lee, E.H. Lee, Y.H. Lee, *Nano Lett.* 9(4), 1401–1405 (2009). DOI: 10.1021/nl803066v.
90. R. Levi et al., *Nano Lett.* 13, 3736–3741 (2013).
91. M. W. Hafiz et al., *ACS Appl. Mater. Interfaces*, 7(42), 23589 (2015) DOI: 10.1021/acsami.5b06825.
92. S. Jo et al., *Nano Lett.* 14, 2019–2025 (2014). DOI: 10.1021/nl500171v.
93. P. Vogt, P. De Padova, C. Quaresima, J. Avila, E. Frantzeskakis, M.C. Asensio, A. Resta, B. Ealet and G. Le Lay, *Phys. Rev. Lett.* 108, 155501 (2012).
94. B. Feng, Z. Ding, S. Meng, Y. Yao, X. He, P. Cheng, L. Chen, K. Wu, *Nano Lett.* 12, 3507–3511 (2012).
95. D. Chiappe, C. Grazianetti, G. Tallarida, M. Fanciulli, A. Molle, *Adv. Mater.* 24, 5088–5093 (2012). DOI: 10.1002/adma.201202100.
96. A. Fleurence, R. Friedlein, T. Ozaki, H. Kawai, Y. Wang, Y. Takamura, *Phys. Rev. Lett.* 108, 245501 (2012).
97. L. Meng et al., *Nano Lett.* 13, 685–690 (2013). DOI: 10.1021/nl304347w.
98. M.E. Davilla, L. Xian, S. Cahangirov, A. Rubio, G. Le Lay, *New J. Phys.* 16, 095002, (2014).
99. S.S. Cahangirov, M. Topsakal, E. Aktürk, H. Sahin, S. Ciraci, *Phys. Rev. Lett.* 102, 236804, (2009).
100. M. Houssa, G. Pourtois, V.V. Afanas'ev, A. Stesmans, *Appl. Phys. Lett.* 97(11), 112106, (2010).

101. Z. Ni et al., *Nano Lett.* 12, 113–118 (2012).
102. H.H. Gurel, V.O. Ozcelik, S. Ciraci, *J. Phys. Condens. Matter* 25(30), 305007, (2013).
103. W.F. Tsai, C.Y. Huang, T.R. Chang, H. Lin, H.T. Jeng, A. Bansil, *Nat. Commun.* 4, 1500 (2013).
104. V. Vargiamidis, P. Vasilopoulos, G.Q. Hai, *J. Phys. Condens. Matter* 26, 345303 (2014).
105. L.C. Lew Yan Voon, E. Sandberg, R.S. Aga, A.A. Farajian, *Appl. Phys. Lett.*, 97, 163114, (2010).
106. M. Houssa, E. Scalise, K. Sankaran, G. Pourtois, V.V. Afanas'ev, A. Stesmans, *Appl. Phys. Lett.* 98(22), 223107 (2011).
107. R. Quhe et al., *Sci. Rep.* 2, 853 (2012).
108. Y. Ding, Y. Wang, *Appl. Phys. Lett.* 100, 083102 (2012).
109. B. van den Broek, M. Houssa, E. Scalise, G. Pourtois, V.V. Afanas'ev, A. Stesmans, *Appl. Surf. Sci.* 291, 104 (2014).
110. T.P. Kaloni, N. Singh, U. Schwingenschlögl, *Phys. Rev. B* 89, 035409 (2014).
111. T.P. Kaloni, Y.C. Cheng, U. Schwingenschlögl, *J. Appl. Phys.* 113, 104305 (2013).
112. H. Zhao, *Phys. Lett. A* 376(46), 3546–3550 (2012).
113. X.S. Ye, Z.G. Shao, H. Zhao, L. Yang, C.L. Wang, *RSC Adv.* 4, 37998–38003 (2014).
114. F. Pan et al. *Sci. Rep.* 5, 9075 (2015).
115. H. Li et al., *Eur. J. Phys. B* 85, 274 (2012).
116. M.A. Kharadi, G.F.A. Malik, K.A. Shah, F.A. Khanday, *IEEE Trans. Electron Devices* 66(11), 4976–4981 (2019). DOI: 10.1109/TED.2019.2942396.

5 Gate and Channel Engineered Nanoscale Electronic Devices

5.1 INTRODUCTION TO NANOSCALE DEVICES

Semiconductors provide unique electrical characteristics varying between quasi-metallic and insulating nature and have become integral part of modern appliances used in our day-to-day life. In fact, semiconductor technology is having unprecedented influence on global economy and cultures worldwide [1]. A semiconductor material, at microscopic level, is characterized by the bands of energy with electrons and holes as carriers of charge [2]. The density of these charge carriers depends on the material and its energy state externally influenced by temperature, doped impurity, and applied voltage [3]. Silicon has been the natural choice as a semiconducting material due to its excellent electrical properties, higher operating temperature, and abundance on earth [4]. Nowadays, several other materials including compound semiconductors, 2D materials, and nanostructures are also being explored for the development of integrated circuits (ICs) [5]. Right from their inception, transistor has been the fundamental building block of ICs.

The transistor in its point contact form was first fabricated by Bardeen and Brattain, and subsequently a three-layer bipolar transistor was realized by Shockley in 1948 [6,7]. The commercialization of bipolar transistor leads to the rise of modern electronic-based solid-state semiconductor technology. The concept of modern day metal oxide semiconductor field effect transistor (MOSFET) utilizing capacitive control to alter the resistance of the semiconductor material-based channel was proposed by Lilienfeld in 1926. However, the correct theory was advanced by Shockley in 1952 [8–10]. Dacey and Ross realized the field-effect transistor (FET) based on Shockley's theory with a metal–semiconductor junction gate in 1953. The performance of this transistor did not offer any significant advantage over bipolar transistor due to the presence of interface traps between SiO_2 and silicon. D. Kahng and M.M. Atalla presented the first silicon–silicon dioxide-based isolated gate FET (modern day MOSFET) in 1963 in which the interface was made almost free of traps [11]. The basic structure of MOSFET (enhancement type) is shown in Figure 5.1. It consists of source and drain regions separated by a semiconducting channel. The conductivity of the semiconducting channel is controlled by potential at the gate electrode separated from the channel by an oxide. If the source and drain are n-type material and the substrate is a p-type material, the MOSFET formed is called n-type MOSFET. On the other hand, the p-channel counterpart of the MOSFET consist of p-type source and drain regions and n-type substrate. The p–n junction formation between

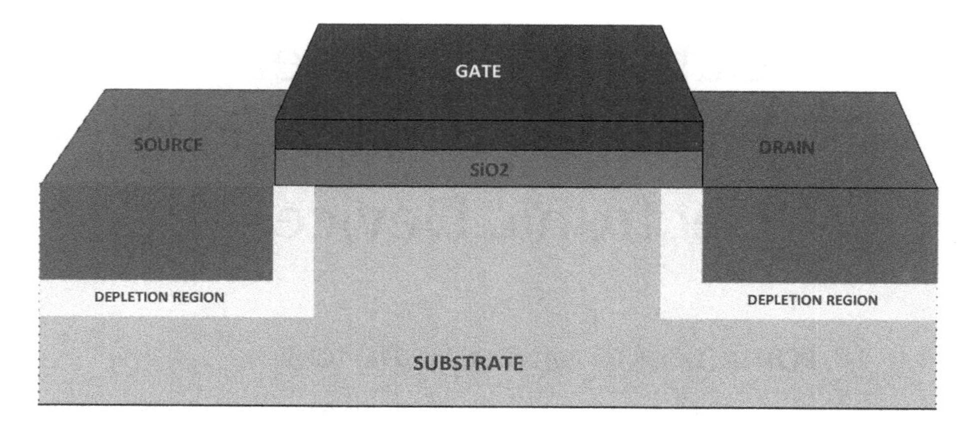

FIGURE 5.1 Structure of conventional MOSFET.

source/drain and substrate can be noted and plays an important role in modern day MOSFETs.

Following the MOSFET realization, the ICs using MOSFETs were developed and the field of complementary metal oxide semiconductor (CMOS) technology came into being, which revolutionized the whole semiconductor industry. CMOS consists of p-type and n-type MOSFETs (P-MOSFET and N-MOSFET) pair, which offered zero power consumption except during a signal transition and an operating region in an amplifier with infinite voltage gain. It offered simple metal-oxide-silicon fabrication process resulting in higher yield and compactness of the device. Subsequently, while working on the realization of ICs at Fairchild, Gordon Moore in 1965 predicted that the complexity of ICs, in terms of the number of elements on each chip, would continue to double each year – this prediction later became famous as Moore's law [12]. The spectrum of ICs progressed from three transistors and one resistor in 1959 to first solid-state microprocessor (Intel 4004) developed in 1971 containing about 2,300 transistors on a single silicon chip [13].

The enhancement in the computational efficiency involves more operations per second, which required more transistors per IC. Stepping away from steady progression of increasing chip size, scaling of individual component feature size was presented to accommodate more transistors per chip to meet the performance and cost predictions. The scaling behavior of MOSFET has been a field of active research to maintain the long channel behavior in the scaled down device. The scaling theory, presented by R. Dennard in 1972, was used as a guide for MOSFET miniaturization in order to avoid deleterious effects and at the same time reduce the circuit delay time [14]. According to this theory, if the device physical dimensions and applied potential are scaled by a common factor called scaling factor, the shape of the electric field pattern within the scaled-down device remains the same as that of the original device. This simple scaling rule enabled the progress of large-scale integration (LSI) towards very-large-scale integration (VLSI) with simultaneous benefits in switching speed and power consumption. The scaling down of MOSFET in submicron scale for VLSI was more aggressive as the number of transistors in a chip was increased by more than the square of scaling factor [15]. This led to modification in the scaling rules to address the limiting factors in a submicron MOSFET like temperature variation of threshold voltage and non-scalability of the junction built-in potential, which leads to a larger depletion region width relative to device dimensions. Baccarani et al. proposed generalized scaling theory which allowed for an independent scaling for physical dimensions and applied voltages and thus paved the way for the realization of sub-0.1-micron (nanometer regime) MOSFET device [16]. However, the scaling of conventional MOSFET in nanometer regime has been less promising due to the degradation of electrical characteristics even if the electric field is not high and small geometry-related issues [17]. The MOS scaling in nanometer regime leads to short channel effects and reliability issues, which have become areas of increasing importance. Short channel effects can be attributed to limitations imposed on carriers in the channel and variations in the threshold voltage due to small device geometry. Hot carriers and direct carrier tunneling through gate-oxide challenge the reliability of sub-100-nm transistors.

5.1.1 Electrostatic Effects

The electric field is inversely proportional to dimension, and in a nanometer regime MOSFET structure, the electric fields are high even if the applied voltages are small. High electric fields can produce non-ideal effects such as gate-oxide breakdown, source-to-drain leakage conduction due to barrier lowering at source end (drain-induced barrier lowering (DIBL)), and avalanche breakdown due to the generation of hot carriers in the channel. The high longitudinal electric fields also lead to the saturation of carrier velocity in the channel and thus make device slow [18–20]. If the longitudinal electric field is further increased, it leads to uncontrolled avalanche breakdown current through the depletion region between source/drain and substrate [21].

5.1.2 Threshold Voltage Roll-Off

In logic applications, the MOSFET is used as a switch, which ideally has zero ON resistance and infinite OFF resistance. Ideally, a switch turns ON if the applied voltage is greater than 0 V but, in case of MOSFET, the ON state is decided by the threshold voltage. If the applied gate source voltage (V_{GS}) is greater than threshold voltage (V_{TH}), the MOSFET turns in ON state and conducts current. The MOSFET remains in the OFF state and does not conduct the current if the applied V_{GS} is less than the V_{TH}. Thus, the threshold voltage plays a critical role in circuit operation, and it is also a very important design parameter for MOSFET due to its dependence on doping in the silicon body, on the depleted charge in the channel due to gate voltage, on the trapped charge at the silicon–silicon dioxide interface and on the gate material [22]. In small geometry MOSFET, part of channel depletion charges is influenced by the presence of source and drain diffusion regions. For sub-100 nm MOSFET, the charge associated with p–n junctions becomes relatively large part of the depletion charge (present in the small channel), and the gate control over the channel electrostatics is reduced. The control is further reduced as the drain–source voltage (V_{DS}) is increased. Thus, in the design of a nanoscale MOSFET, the threshold voltage sensitivity to channel length and drain voltage is a crucial issue.

5.1.3 Leakage Currents

Scaling down of MOSFET in nanometer regime also poses the challenge of increased OFF-state leakage current. The OFF-state leakage current is an important device parameter as it determines the stand-by power dissipation in the MOS-based circuits. The various leakage currents associated with MOSFETs are discussed next.

5.1.3.1 Gate Leakage Current

Silicon dioxide (SiO_2) used as a gate-oxide to isolate the gate electrode from the silicon body has been a defining feature of the MOSFET technology. It provides excellent dielectric properties like large band gap of about 9 eV and fabrication stability with silicon body [23]. Gate leakage current due to miniaturization of the oxide in terms of its thickness even less than 20 Å has become a significant issue in nanoscale MOSFET. It involves the direct tunneling of charge carriers through the oxide from channel to the gate or vice versa even for low voltage across the oxide.

5.1.3.2 Subthreshold Leakage Current

Subthreshold leakage current is due to the diffusion of minority carriers in the channel in a weak inversion region ($V_{GS} < V_{TH}$) of operation. The intensity of subthreshold current shows the penetration of electric field from source and drain. In nanoscale MOSFET, since the amount of depletion charge controlled by this lateral electric field is large, it leads to higher subthreshold drain current. The slope of this current gives subthreshold swing, and degradation of subthreshold swing is one of the major problems in nanometer MOSFET [24].

5.1.3.3 Junction Leakage Current

It consists of two currents: (1) minority carrier diffusion/drift near the edge of the depletion region and (2) electron–hole pair generation in the depletion region. If both n- and p-regions are heavily doped, band-to-band tunneling dominates junction leakage. The initial current is the reverse current of the p–n junction diodes which is very small but increases as the collision in the depletion region increase (due to high lateral field) leading to even avalanche breakdown.

5.2 NON-CONVENTIONAL SOLUTIONS TO MINIATURIZATION PROBLEMS

5.2.1 Silicon-On-Insulator

The channel formation in a MOSFET takes place in few micrometers below the oxide. The thickness of substrate is kept much greater than that to ensure the structural feasibility of the device. However, this unused substrate adds several parasitics and contributes to leakage currents directly from source and drain regions [25]. In addition, the unused substrate reduces the mobility of the charge carriers. These effects aggravate especially with higher body doping and small dimensions. To maintain the structural feasibility and eliminate these effects, an insulating layer is added within the body of substrate as shown in Figure 5.2. Such a technique is called

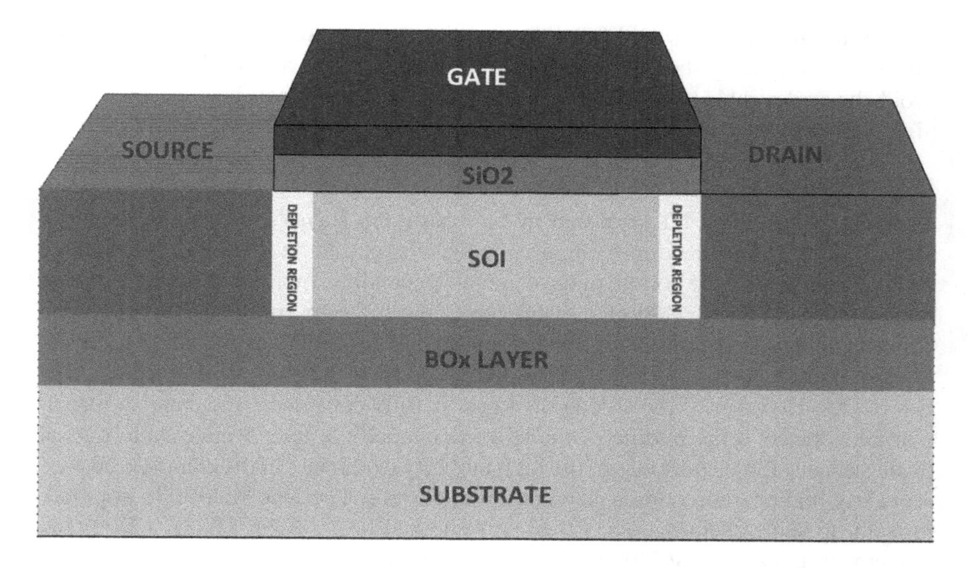

FIGURE 5.2 SOI MOSFET device structure.

silicon-on-insulator (SOI) technology. Depending upon whether the depletion region of the channel consumes the silicon depth between the oxides partially or fully, the SOI MOSFETs are categorized as partially depleted SOI (PD-SOI) MOSFETs and fully depleted SOI (FD-SOI) MOSFETs. The buried oxide (BOx) layer within the substrate offers low parasitics, removes the leakage paths, and enables the high speed of operation. Besides, SOI platform offers enhanced channel mobility, avoids device latch-up, reduces random dopant fluctuation, and decreases variability in threshold voltage [26,27]. However, the main problem in SOI is the self-heating and therefore prone to heat accumulation as the thermal conductivity of SiO_2 is about two orders of magnitude less than silicon [28]. Replacing SiO_2 with air (silicon-on-nothing (SON)) has been proposed as an alternate solution for advanced scalability due to better heat dissipation and good control of the fringing fields [29,30].

5.2.2 Multigate MOSFET

The MOSFET has been scaled into nanoregime to maintain the performance predicted by International Technology Roadmap for Semiconductors (ITRS) while at the same time achieving ultra-large-scale-integration (ULSI) implementation [31]. To avoid the non-ideal effects and electrostatic limitations, several techniques have been proposed by the researchers. Besides, in order to increase the control of gate over the channel, thin oxides or high-k dielectrics were proposed. However, the problem with thin oxides or high-k dielectrics is their structural instability and difficulty of fabrication [32]. Colinge in 2004 pointed out that electrostatic control of gate over the channel can be enhanced by using multiple gates while at the same time using thicker oxides [33]. Based on the number of gates used, the multigate MOSFETs are divided into double-gate (DG), trigate (TG), and surrounding gate (SG) or gate-all-around (GAA) MOSFETs. The multigate transistors are considered as the future for mitigating the problems of conventional nanoscale MOSFETs [34]. To avoid the undesirable short channel effects, multiple gates can be used instead of high doping in the channel.

5.2.2.1 Double-Gate (DG) MOSFET

The planar DG MOSFET proposed by Sekigawa and Hayashi in 1984 is shown in Figure 5.3 [35]. As shown in the figure, one gate is placed on the top of the channel and the other gate at the bottom of the channel. Depending upon whether the depletion regions formed by top and bottom gates consumed full or partial thickness of silicon between oxides, the MOSFET is divided into FDDG MOSFET and PDDG MOSFET. In case, the full silicon thickness is not consumed, two channels are formed near the two oxides. However, if the silicon thickness is fully consumed, the channel formation takes place at the middle of the silicon channel and thus avoids scatterings at oxide surfaces [36]. This leads to the high mobility of carriers in the channel and consequently higher transconductance and drive current. The DG MOSFETs are more resistant to short channel effects than conventional bulk MOSFETs [37]. However, implementing DG MOSFETs is difficult as maintaining double-gate symmetry is a serious technology challenge. Therefore, a nonplanar DG MOSFET device called FinFET was proposed which promises performance and scalability in nanoregime and at the same time eliminates the technology barrier of planar DG MOSFET [38].

5.2.2.2 Trigate (TG) MOSFET

TG MOSFET was first proposed by Doyle et al. in 2003. The channel is surrounded by gate from three sides [39]. The SOI version of TG MOSFET was reported by Colinge and is shown in Figure 5.4 [40]. Unlike the DG MOSFETs, the TG MOSFET does face the problem of gate symmetry and has better current drive than DG MOSFET

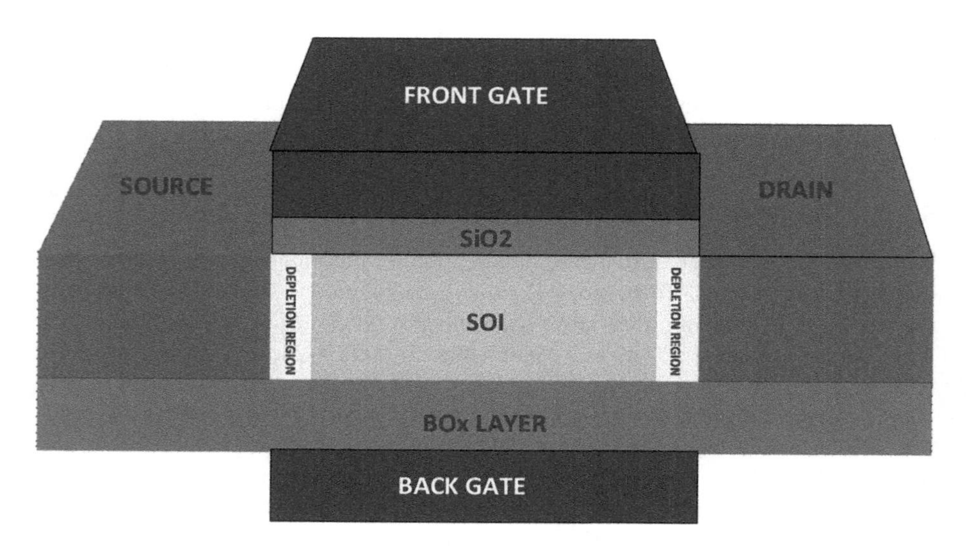

FIGURE 5.3 Double-gate SOI MOSFET structure.

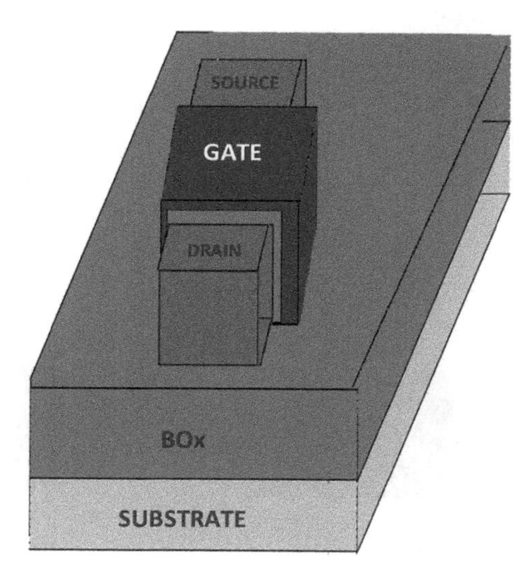

FIGURE 5.4 Trigate SOI MOSFET structure.

due to the formation three conduction channels. Besides, it offers better immunity to short channel effects than DG MOSFET [41]. To avoid the corner effects, the triangular and trapezoidal versions of TG MOSFETs have also been reported [42].

5.2.2.3 Gate-All-Around (GAA) MOSFET

The channel GAA MOSFET is surrounded by gates from all sides. Based on their shapes, GAA MOSFETs are categorized as square-, rectangular-, and circular-shaped GAA MOSFETs as shown in Figure 5.5 [43–45]. GAA MOSFETs provide better scalability and avoid short channel effects. Since the channel is surrounded by gates from all sides, the GAA MOSFET provides ideal subthreshold swing [46]. If the thickness of the channel is of few atoms, then rectangular- and circular-shaped GAA transistors are called nanosheet (NS) MOSFET and nanowire (NW) MOSFET, respectively.

5.3 GATE AND CHANNEL ENGINEERING TECHNIQUES

5.3.1 Gate-Oxide Stack

In order to increase the electrostatic control of the gate over the channel, the oxide thickness is reduced. However, the reduced oxide thickness leads to more tunneling and thereby increases the leakage current between gate and channel and consequently increases the static power consumption of the device [47]. One of the solutions to mitigate this problem is to use multiple gates as discussed earlier. Using high-k dielectric instead of SiO_2 can also be a solution. However, the high-k dielectric results in increased interface defects. Another solution is gate-oxide stack [48]. In this technique, a thick layer of high-k oxide is sandwiched with a thin layer of SiO_2. The thin layer of high-k oxide results in increased oxide thickness, thereby reducing tunneling/leakage current while at the same time providing more

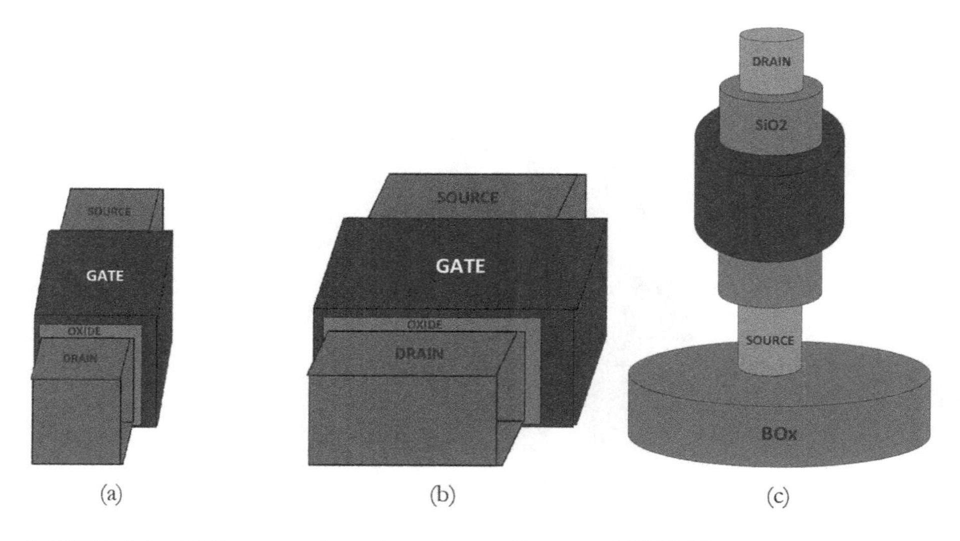

FIGURE 5.5 Different configurations of gate-all-around MOSFET structure. (a) Square-shaped GAA or quad gate MOSFET. (b) Rectangular-shaped GAA MOSFET. (c) Cylindrical-shaped GAA MOSFET.

electrostatic control of the gate over the channel. The gate-oxide stack in multigate MOSFETs enhances the immunity against short channel effects, increases ON/OFF current ratio, increases intrinsic DC gain (A_V), cutoff frequency (f_T), and maximum oscillation frequency (f_{MAX}) [49,50]. The technique however needs an extra fabrication step which makes it bit costlier.

5.3.2 Gate Metal Work Function Engineering

In order to increase the mobility of carriers in nanoscale MOSFETs, the channel is either lightly doped or undoped [51]. The doping of the channel controls the threshold voltage of the MOSFET and immunity to short channel effects. However, if undoped channel is used, controlling the threshold voltage is not possible. In that case, the gate material will control the threshold voltage of the device. When using the traditional polysilicon as the gate material in nanoscale MOSFET, it forms a depletion region with the high-k dielectric, thereby increasing the effective oxide thickness [52]. This problem can be avoided by using highly doped polysilicon gates. However, the highly doped polysilicon gates lead to negative threshold voltage in n-channel MOSFETs and positive threshold voltage in p-channel MOSFETs. These effects make polysilicon unsuitable as gate material for nanoscale MOSFET, which can be avoided by using metal gates. Metal gates are compatible with high-k dielectrics and provide more carrier mobility in the channel by reducing the transverse electric field [53,54]. Besides, nowadays the work function of the metal gate can be tuned to adjust the threshold voltage [55,56]. It has also been demonstrated that using multiple materials with different work function in gate leads to the enhanced mobility of the carriers in the channel and provides a screening effect to suppress short channel effects.

For example, if two metals of different work functions are used in gate with metal of higher work function placed on the source side and the metal with lower work function placed on the drain side, the mobility of the carriers in the channel enhances and a screening effect to suppress short channel effects is produced. This becomes possible due to the fact that using the work functions in the above manner results in higher threshold voltage towards source side and less threshold voltage towards drain side [57]. The same effect has been observed in multigate devices as well [58–63].

5.3.3 Channel Engineering

Traditionally, the doping of the channel is kept uniform and the whole channel is inverted by the gate potential. However, if the graded doping is considered instead of uniform doping, the region with less doping inverts first and thereby decreases the effective length of the device and hence increases the drive current of the device. For example, if the drain end of the channel is lightly doped or undoped and source side of the channel is heavily doped, the drain side of the channel will have lower threshold voltage and source side of the channel will have higher threshold voltage. As the gate potential is applied, the drain side of the channel becomes an extension of drain and thereby reduces the effective channel length and increases the drive current of the device. The low doping on the drain side also increases the mobility of carriers in the channel. Graded channel devices have improved punch-through and DIBL characteristics while at the same time enhancing the device reliability. Graded channel devices do not require complex fabrication processes and are fully compatible with mainstream CMOS technology [64,65].

5.3.4 Strained Layer

If a thin layer of a material is epitaxially grown on the thick substrate of the same material, no mismatch occurs and no dislocations are formed. However, if the lattice constant of the top thin layer is different from the substrate, the lattice constant of the top thin layer changes to match the thick substrate or the strain of the top layer takes place [66]. Depending upon whether the original lattice constant of the top thin layer was greater or less than the lattice constant of the substrate, compressive or tensile strain of the top layer takes place [67]. The maximum thickness (critical thickness) of the top strained layer depends on the difference in the original lattice constants of the top layer and substrate. If the thickness of top layer exceeds the critical thickness, strain is relieved and dislocations are formed. The strain leads to the change in energy levels and thereby energy gap. The strain usually lowers the effective mass of the holes and thereby increases the mobility of hole [68]. Therefore, strained layers are used in the channel of MOSFET to achieve higher mobility of carriers and higher drive current.

5.4 MULTIGATE MULTI-MATERIAL MOSFET

As discussed earlier, based on the number of gates, the number of materials used in gates, and channel engineering, several combinations are possible. Here, the structure and performance of dual-material double-gate (DMDG) SON MOSFET has been considered. The schematic structure of high-k dielectric DMDG SON MOSFET is shown in Figure 5.6 [69]. Vacuum or air has been considered as the buried dielectric in the device. The work function (φ_{m1}.) of M_1 is greater than work function (φ_{m2}) of M_2.

The potential distribution $\psi_1(x, y)$ and $\psi_2(x, y)$ corresponding to gate material M_1 (with work function φ_{m1}) in region I and M_2 (with work function φ_{m2}) in region II, respectively, can be derived by solving 2D Poisson's equation and using Gauss's law to obtain the boundary conditions [69]. The obtained 2D potential distribution due to dual-material gate can be expressed as

$$\psi_1(x, y) = \psi_{s1}(x) + \frac{\varepsilon_{ox}}{\varepsilon_{Si} t_{ox}} \left(\psi_{s1}(x) - V'_{gs1}\right) y + \frac{\left(V'_{gs1} - \psi_{s1}(x)\right)\left(1 + \frac{C_{ox}}{C_{si}} + \frac{C_{ox}}{C_{\text{box}}}\right)}{t_{si}^2 \left(1 + 2\frac{C_{si}}{C_{\text{box}}}\right)} \tag{5.1}$$

And

$$\psi_2(x, y) = \psi_{s2}(x) + \frac{\varepsilon_{ox}}{\varepsilon_{Si} t_{ox}} \left(\psi_{s2}(x) - V'_{gs2}\right) y + \frac{\left(V'_{gs2} - \psi_{s2}(x)\right)\left(1 + \frac{C_{ox}}{C_{si}} + \frac{C_{ox}}{C_{\text{box}}}\right)}{t_{si}^2 \left(1 + 2\frac{C_{si}}{C_{\text{box}}}\right)} \tag{5.2}$$

N_a is the uniform channel dopant concentration, ε_{Si} is the dielectric constant of silicon, ε_{ox} is the dielectric constant of oxide, ε_{box} is the dielectric constant of Box layer, C_{Si} is the capacitance of silicon, C_{ox} is the capacitance of oxide, C_{box} is the

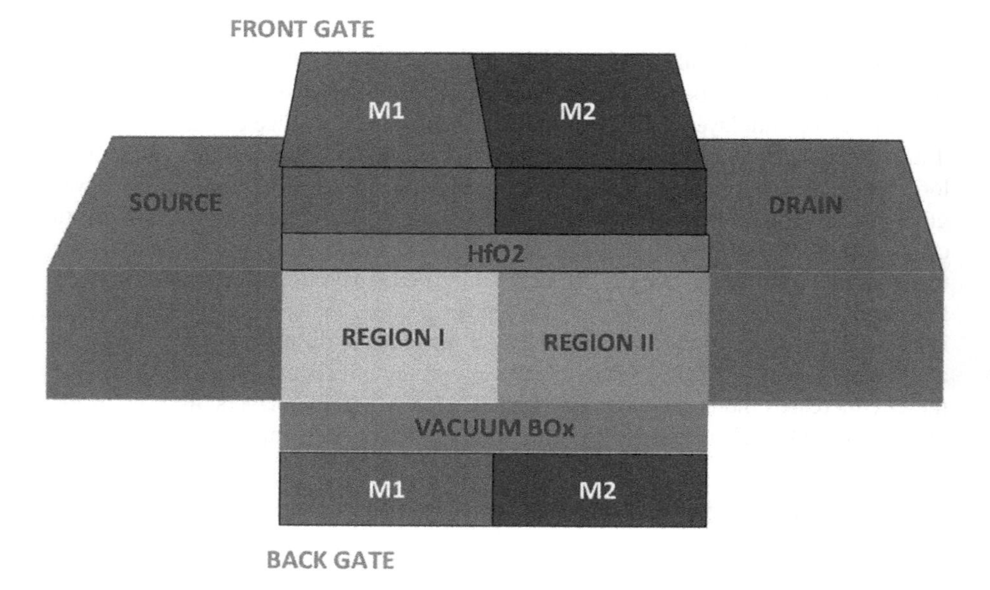

FIGURE 5.6 Schematic of DMDG SON MOSFET with high-*k* front oxide.

capacitance of Box layer, t_{Si} is the thickness of silicon, and t_{ox} is the thickness of oxide; $V'_{gs1} = V_{gs} - V_{Fb1}$ and $V'_{gs2} = V_{gs} - V_{Fb2}$, where V_{Fb1} and V_{Fb2} are flat-band voltages corresponding to gate materials M_1 and M_2, respectively.

The surface potential distribution functions in terms of device parameters and bias voltage can be obtained as [69]

$$\psi_{s1}(x) = -\sigma_1 + \left[\frac{\{(V_{bi} + V_{DS} + \sigma_2) - (V_{bi} + \sigma_1)\exp(-\Gamma) + (\sigma_1 - \sigma_2)\eta L_2)\}\exp(-\Gamma)}{1 - \exp(-2\Gamma)}\right] \times \exp\left(-(\alpha)^{1/2} x\right) + \left[\frac{(V_{bi} + \sigma_1)e^{\Gamma} - (V_{bi} + V_{DS} + \sigma_2) + (\sigma_2 - \sigma_1)\cos h(\eta L_2)}{\exp(\Gamma) - \exp(-\Gamma)}\right] \times \exp\left((\alpha)^{1/2} x\right) \quad (5.3)$$

And

$$\psi_{s2}(x) = -\sigma_2 + \left[\frac{\{(V_{bi} + V_{DS} + \sigma_2) - (V_{bi} + \sigma_1)\exp(-\Gamma) + (\sigma_1 - \sigma_2)\cos h(\eta L_2)\}\exp(-\Gamma)}{1 - \exp(-2\Gamma)}\right] \times \exp\left((\alpha)^{1/2} x\right) + \left[\frac{(V_{bi} + \sigma_1)e^{\Gamma} - (V_{bi} + V_{DS} + \sigma_2) + (\sigma_2 - \sigma_1)\cos h(\eta L_2)}{\exp(\Gamma) - \exp(-\Gamma)}\right] \times \exp\left(-(\alpha)^{1/2} x\right) + \frac{(\sigma_2 - \sigma_1)}{2}\cos h\left((\alpha)^{1/2}(x - L_1)\right) \quad (5.4)$$

where $\sigma_i = \frac{\xi_i}{\lambda}, \Gamma = \left(L(\alpha)^{\frac{1}{2}}\right)$, L_1 is the length of region I, and L_2 is the length of region II.

The threshold voltage is taken to be that value of gate-to-source voltage (V_{GS}) for which $\psi_{s,\min} = 2\varphi_F$, where φ_F is the difference between the extrinsic Fermi level in the bulk region and the intrinsic Fermi level and is given by

$$\varphi_F = \frac{KT}{q}\ln\left(\frac{Na}{ni}\right) \quad (5.5)$$

Since the work function of M_1 is greater than that of M_2, $\psi_{s,\min}$ occurs under gate material M_1 and can be found by solving

$$\left.\frac{\partial^2 \psi_{S1}(x)}{\partial x^2}\right|_{x = x_{\min}} = 0 \quad (5.6)$$

For simulations purposes, the two materials M_1 and M_2 are considered of equal lengths, i.e., $L_1 = L_2 = L/2$ with work functions of 4.8 and 4.6 eV, respectively. The analysis has been carried out using the parameters given in Table 5.1.

The surface potential distribution of single-material $(\phi_m = 4.8\,eV)$ double-gate (SMDG) and DMDG SON MOSFET with high-k dielectric has been compared as a function of position along the channel length shown in Figure 5.7 for the given device parameters and stated bias voltages. It can be observed that the potential profile in SMDG is smooth while the DMDG has step potential profile. The step potential shields the position of minimum channel potential from the fluctuations in V_{DS}. Moreover, the position of minimum channel potential in DMDG is shifted towards the source end compared to SMDG, which indicates device threshold voltage to have higher immunity against V_{DS}-induced reduction. Figure 5.8 shows the variation of surface potential along the channel length of DMDG with three different HfO_2 oxide thicknesses: $t_{ox} = 0.1$, 0.2, and 0.3 nm. From the graph, it is obvious that the step profile tends to flatten with decreasing oxide thicknesses indicating higher gate control over the channel electrostatics and immunity against SCEs. Moreover, the position of minimum channel potential is lowered for thinner gate-oxides implying the higher threshold

TABLE 5.1
Device Parameters

ϕ_{m1} (eV)	ϕ_{m2} (eV)	t_{ox} (nm)	t_{si} (nm)	t_{box} (nm)	L (nm)	N_a (cm^{-1})	N_d (cm^{-1})
4.8	4.6	2	5	3	40	10^{21}	5×10^{26}

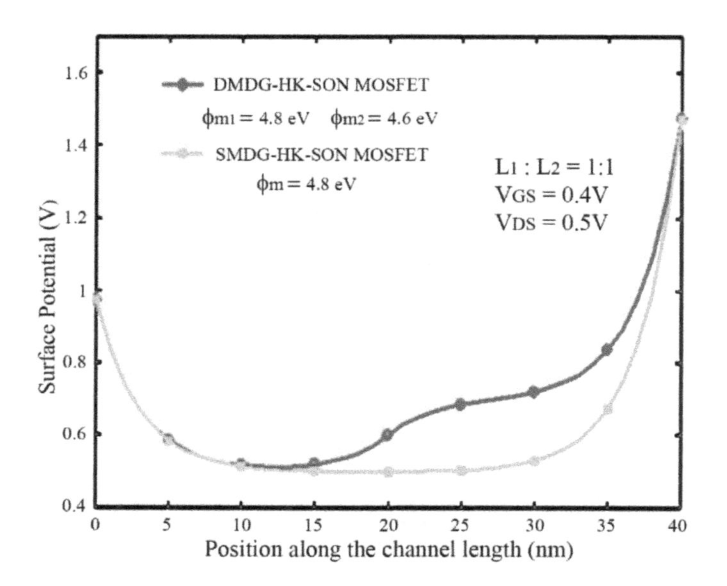

FIGURE 5.7 Surface potential variations along the channel for SMDG and DMDG high-*k* SON MOSFET.

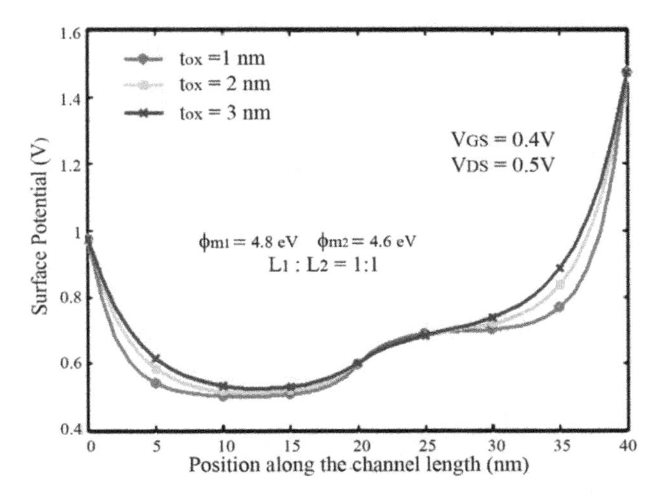

FIGURE 5.8 Surface potential variations along the channel for DMDG SON MOSFET with high-*k* gate dielectric for different oxide thicknesses.

voltage due to enhanced gate control. The effect of silicon channel thickness (t_{si}) on the surface potential has also been plotted as a function of channel length shown in Figure 5.9 for a channel thickness of 5, 10, and 15 nm. From the graph, it is clear that as the channel thickness decreases, the step channel potential profile in high-*k* DMDG-SON-MOSFET improves suggesting higher channel control for thinner body devices. Figure 5.10 shows the variation of surface electric field as a function of channel length with different channel thicknesses. The step increase in the surface potential consequently leads to an additional electric field peak at the interface of dual gate materials in addition to the electric field peaks at the source and drain ends. The higher electric

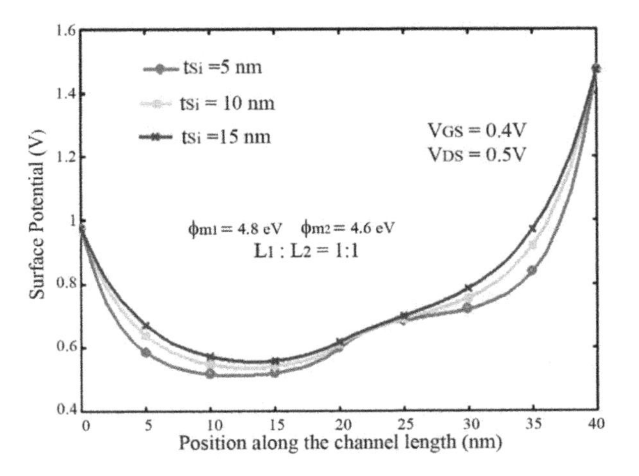

FIGURE 5.9 Surface potential variations along the channel length for DMDG SON MOSFET with high-*k* gate dielectric for different channel thickness.

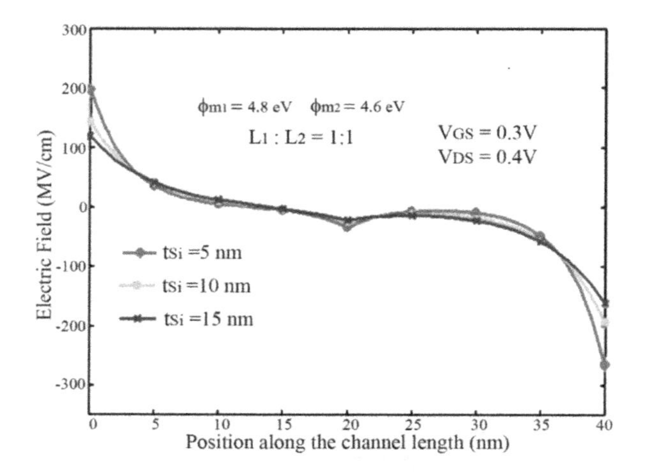

FIGURE 5.10 Electric field variation along the channel for DMDG SON MOSFET with high-*k* gate dielectric for different channel thickness.

field in the channel increases the electron velocity resulting in increased drain current. However, the thinner body also leads to higher drain side electric field, which may give rise to hot carriers. The effects of using low-*k* and high-*k* gate dielectrics on the electric field are shown in Figure 5.11 as a function of position along the channel length. The figure shows that the electric field is higher when using HfO_2 as gate-oxide compared to SiO_2 as the higher dielectric constant oxide will induce higher gate capacitance. This higher gate capacitance will not only improve gate control over the channel but give rise to higher electric field in the channel. The line graphs in the figures represent the analytical model plotted in MATLAB® software, whereas the symbols indicate the simulation results carried out in ATLAS software [70].

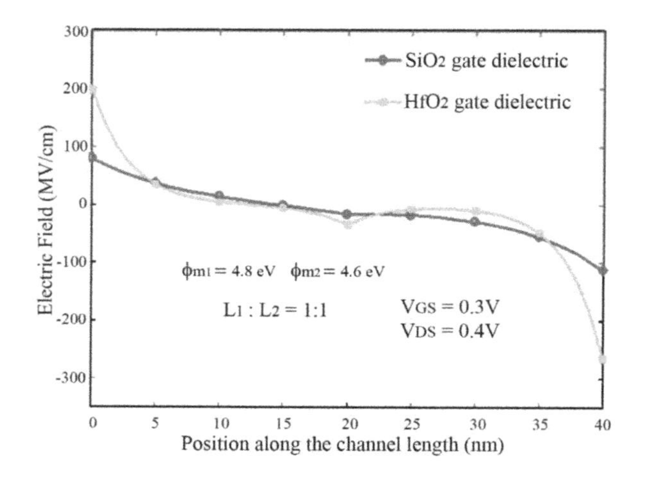

FIGURE 5.11 Electric field variation along the channel for low-*k* and high-*k* gate dielectric materials. Line represents the model, whereas symbols represent simulation results.

5.5 MULTIGATE MULTI-MATERIAL TUNNEL FET

Tunnel field-effect transistor (TFET) with subthreshold swing less than 60 mV/dec and CMOS process compatibility had been considered as a promising device for various low-power and low-voltage applications [71]. Silicon-based TFET suffers from low ion due to poor tunneling probability of carriers from valence band of P+-doped source into the conduction band of intrinsic channel region [72]. To overcome problems of low ion and high ambipolar current in TFET, various devices have been proposed incorporating narrow bandgap materials and structural modifications [73]. Dual-material trigate TFET has been reported in [74] where the gate consists of two materials with low work function material close to source and high work function material towards drain. In this section, the structure and performance of gate engineered trapezoidal trigate TFET has been considered. The schematic of gate engineered trapezoidal trigate TFET is shown in Figure 5.12. The structure consists of silicon buffer/substrate and buried oxide layer of SiO_2 separating substrate and active channel region. The hetero material gate and gate-oxide surround the channel from two lateral sides in addition to top side. The gate is composed of two types of materials with work functions $\varnothing_{m1} = 4.0$ eV, $\varnothing_{m2} = 4.4$ eV, and $\varnothing_{m3} = 4.0$ eV. The gate materials M_1 and M_3 have been considered to have the same work function unless otherwise stated. The device channel region is divided into three regions corresponding to material positions such that $L_1{:}L_2{:}L_3 = 1{:}1{:}1$. L_1, L_2, and L_3 represent the region lengths corresponding to heterogate materials. In this structure, the source region is doped with $N_a = 10^{20}/\text{cm}^3$, drain region is doped with $N_d = 10^{19}/\text{cm}^3$, and the channel is lightly doped with $N_a = 10^{15}/\text{cm}^3$, while source/channel and drain/channel junctions are considered abrupt.

The device parameters used in deriving the model and simulation results are channel length, $L = 30$ nm, thickness of buried oxide layer, $t_{box} = 40$ nm, thickness of front gate-oxide (SiO_2), $t_{ox} = 1$ nm, and thickness of channel, $t_{Si} = 10$ nm. Since the width of the device varies from top of the channel to the bottom as $-\frac{w_1}{2}$ to $+\frac{w_1}{2}$ and $-\frac{w_2}{2}$ to $+\frac{w_2}{2}$ on the z-axis, the inclination of side gates of 0°, 5°, and 10° have been considered with $W_{top} = 10$ nm.

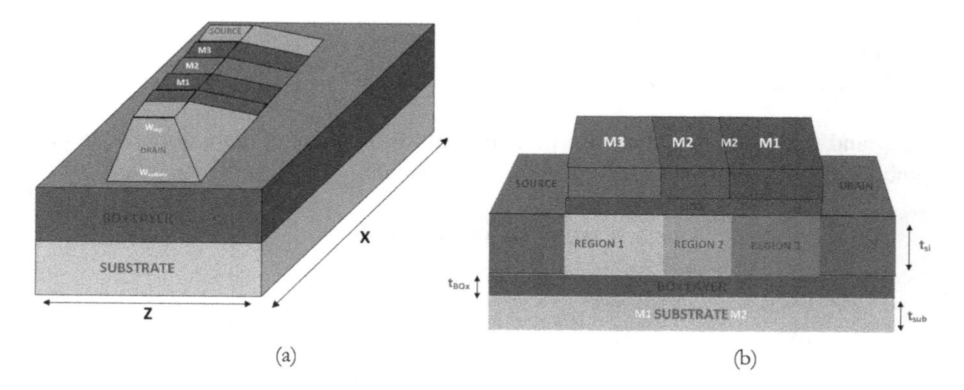

FIGURE 5.12 Three-dimensional schematic and x-y sectional view of gate engineered trapezoidal trigate TFET.

The three channel regions are defined as follows:
Channel region 1

$$\left(0 \le x \le L_1,\ 0 \le y \le t_{si},\ -\frac{w}{2} \le z \le \frac{w}{2}\right)$$

Channel region 2

$$\left(L_1 \le x \le L_2,\ 0 \le y \le t_{si},\ -\frac{w}{2} \le z \le \frac{w}{2}\right)$$

Channel region 3

$$\left(L_2 \le x \le L,\ 0 \le y \le t_{si},\ -\frac{w}{2} \le z \le \frac{w}{2}\right)$$

where w is the equivalent device width and is calculated at the orthocenter of device [75] as

$$w = w_{\text{top}} + \frac{r}{r+1}\left(w_{\text{bottom}} - w_{\text{top}}\right) \tag{5.7}$$

and r is the geometrical factor given by

$$r = \frac{2w_{\text{bottom}} + w_{\text{top}}}{2w_{\text{top}} + w_{\text{bottom}}} \tag{5.8}$$

The potential in the channel follows parabolic approximation [76] and can be expressed as the second-order polynomial function:

$$\psi_k(x,\ y,\ z) = \xi_{0k}(x,\ y)z^2 + \xi_{1k}(x,\ y)z + \xi_{2k} \tag{5.9}$$

where $k = 1,2,3$ and represents three channel regions. The coefficients of Equation (5.9) can be determined by using the following boundary conditions [77]:

1. $\psi_1(0,y,z) = V_{bi,p}$ (5.10)
2. $\psi_3(L,y,z) = V_{bi,n} + V_{DS}$ (5.11)

$V_{bi,\,n}$ and $V_{bi,\,p}$ are the built-in potential at drain/channel and source/channel junctions, respectively, given by $V_{bi,\,n} = v_t \ln\left(N_{d,\text{ drain}} \times N_{a,\text{ channel}} / n_i^2\right)$ and $V_{bi,\,p} = -v_t \ln\left(N_{a,\text{ source}} / N_{a,\text{ channel}}\right)$.

Due to continuity in the flux at the gate-oxide and silicon channel boundary, we have

3. $\left.\dfrac{d\psi_k(x,y,z)}{dy}\right|_{y=0,z=0} = \dfrac{\varepsilon_{ox}}{\varepsilon_{si}t_{ox}}\left(\psi_{sk}(x) - V'_{GSk}\right)$ (5.12)

4. $\left.\dfrac{d\psi_j(x,y,z)}{dz}\right|_{y=t_{si},z=0} = \dfrac{\varepsilon_{ox}}{\varepsilon_{si}t_{\text{box}}}\left(V'_{\text{sub}} - \psi_{bk}(x)\right)$ (5.13)

Because of symmetry about the orthocenter ($z = 0$) along the z-direction, we have

$$\psi_k\left(x, y, -\frac{w}{2}\right) = \psi_k\left(x, y, +\frac{w}{2}\right) \tag{5.14}$$

where ε_{ox} and ε_{si} are the dielectric constants of oxide and silicon, respectively, and t_{ox} and t_{box} are the thicknesses of gate-oxide and BOX layer, respectively. $V'_{GSk} = V_{GS} - V_{FBk}$, where V_{FBk} represents channel flatband voltage of region 1, region 2, and region 3 under M_1, M_2, and M_3, respectively. $V'_{\text{sub}} = V_{\text{sub}} - V_{Fb,\, b}$, where V_{sub} is the substrate bias, and $V_{Fb,\, b}$ is the back channel interface flatband voltage. $\psi_{sk}(x)$ and $\psi_{bk}(x)$ represent the surface potentials at front gate-oxide–channel region interfaces and BOx layer–channel region interface.

Using the above boundary conditions, the coefficients are obtained as

$$\xi_{0k}(x, y, z) = \left(\frac{4}{w^2}\right)\left(\psi_{sk}(x) - \xi_{2k}(x, y)\right) \tag{5.15}$$

$$\xi_{1k}(x,\, y,\, z) = 0 \tag{5.16}$$

The coefficient $\xi_{2k}(x,\, z)$ can be derived by assuming the parabolic potential profile in vertical direction for low Vds such that

$$\xi_{2k}(x, y) = \psi_{sk}(x) + \frac{\varepsilon_{ox}}{\varepsilon_{si} t_{ox}}\left(\psi_{sk}(x) - V'_{GSk}\right) y$$

$$- \frac{\left[\left(\frac{t_{ox}}{t_{\text{box}}} + \frac{C_{\text{box}}}{C_{si}} + 1\right)\psi_{sk}(x) - \left(1 + \frac{C_{\text{box}}}{C_{si}}\right) V'_{GSk}\right]\frac{t_{\text{box}}}{t_{ox}} - V'_b}{t_{si}^2\left(1 + 2\frac{C_{si}}{C_{\text{box}}}\right)} y^2 \tag{5.17}$$

where C_{box} and C_{si} represent the buried oxide capacitance and Si channel capacitance, respectively.

The 3D potential distribution function $\psi_k(x,\, y,\, z)$ for gate engineered trapezoidal trigate TFET can be calculated by solving three-dimensional Poisson's equation [78]:

$$\frac{\partial^2 \psi_k(x,y,z)}{\partial x^2} + \frac{\partial^2 \psi_k(x,y,z)}{\partial y^2} + \frac{\partial^2 \psi_k(x,y,z)}{\partial z^2} = \frac{qN_c(x,y,z)}{\varepsilon_{si}} \tag{5.18}$$

q is electron charge, $N_c(x,\, y,\, z) = \left(N_{a,\, c} + n(x,\, y,\, z)\right)$ where $N_{a,\, c}$ is channel doping concentration and $n(x,\, y,\, z)$ is mobile electron charge density. For full depletion approximation of the channel under zero bias condition, it can be considered that $N_{a,\, c} >> n(x,\, y,\, z)$ and $N_c(x,\, y,\, z) \approx qN_a$.

By solving Equations (5.9) and (5.18) using the coefficients derived in Equations (5.15), (5.16), and (5.17), we get

$$\frac{\partial^2 \psi_{sk}(x)}{\partial x^2} - \alpha \psi_{sk}(x) = \beta_k - 2\frac{C_{si}}{\delta t_{si}}\left[w^2 + 4\left(y^2 + z^2\right)\right]V_b' \tag{5.19}$$

$$\alpha = \left\{ 8\frac{C_{ox}}{\delta}\left[t_{si}^2 \left(1 + \frac{C_{si}}{C_{box}}\right)\right] y + 2\frac{C_{si}}{t_{si}}\left[\frac{t_{box}}{t_{ox}} + \frac{C_{ox}}{C_{si}} + 1\right]\left(w^2 - 4\left(y^2 + z^2\right)\right)\right\}$$

$$\beta_k = qN_{a,c}w^2\delta^{-1}\left(1 + 2\frac{C_{si}}{C_{box}}\right)t_{si}^2 + 4t_{si}^2\frac{\varepsilon_{ox}}{t_{ox}\delta}\left(1 + 2\frac{C_{si}}{C_{box}}\right)yt_{si}^2 V_{GSk}'$$

$$+\left(\frac{t_{box}}{t_{ox}} + \frac{C_{ox}}{C_{si}}\right)\left(w^2 - \left(x^2 + y^2\right)\right)\frac{C_{si}}{t_{si}\delta}V_{GSk}'$$

$$\delta = \left(t_{si}^2\left(1 + 2\frac{C_{si}}{C_{box}}\right)\right)\left(w^2\varepsilon_{si} + C_{ox}\left(w^2 - 4z^2\right)y\right) + \left(\frac{t_{box}}{t_{ox}} + \frac{C_{ox}}{C_{si}} + 1\right)\left(\varepsilon_{si}\left(4z^2 - w^2\right)y^2\right)$$

The general form of surface potential distribution function $\psi_{sk}(x)\big|_{k=1,2,3}$ in gate engineered trapezoidal trigate TFET can be obtained by solving the second-order differential equation obtained in Equation (5.19). Its solution is given as

$$\psi_{sk}(x) = R_k e^{\eta X_k} + S_k e^{-\eta X_k} - \sigma_k \tag{5.20}$$

where $\sigma_k = \beta_k / \alpha$, and X_k takes the value $X_1 = x$, $X_2 = x - L_1$, and $X_3 = x - (L_1 + L_2)$ in the channel regions 1, 2, and 3, respectively. R_k and S_k are arbitrary constants to be determined by using the following boundary conditions:

1. $\psi_1(L_1,0,0) = \psi_2(L_1,0,0)$ (5.21)

2. $\psi_{s1}(L_1) = \psi_{s2}(L_2)$ (5.22)

3. $\left.\frac{\partial \psi_{s1}(x,y,z)}{\partial x}\right|_{x=L_1} = \left.\frac{\partial \psi_{s2}(x,y,z)}{\partial x}\right|_{x=L_1}$ (5.23)

4. $\psi_2(L_1 + L_2,0,0) = \psi_3(L_1 + L_2,0,0)$ (5.24)

5. $\psi_{s2}(L_1 + L_2) = \psi_{s3}(L_1 + L_2)$ (5.25)

6. $\left.\frac{\partial \psi_{s2}(x,y,z)}{\partial x}\right|_{x=L_1+L_2} = \left.\frac{\partial \psi_{s3}(x,y,z)}{\partial x}\right|_{x=L_1+L_2}$ (5.26)

Using the boundary conditions given in Equations (5.21)–(5.26), the coefficients R_k and S_k are obtained as

$$R_1 = \left\{\left(V_{bi,n} + V_{DS} + \sigma_3\right) - \left(V_{bi,p} - \sigma_1\right)e^{-\eta L} + (\sigma_1 - \sigma_2)\cosh\left(\eta\left(L_2 + L_3\right)\right)\right.$$

$$\left. + (\sigma_2 - \sigma_3)\cosh(\eta L_3)\right\}\left\{\sinh^2(\eta L_3)\right\}^{-1}$$

$$S_1 = \left(V_{bi,n} - \sigma_1\right) - A_1$$

$$R_2 = A_1 e^{\eta L_1} - (\sigma_1 - \sigma_2)/2$$

$$S_2 = B_1 e^{-\eta L_1} - (\sigma_1 - \sigma_2)/2$$

$$R_3 = A_2 e^{\eta L_2} - (\sigma_2 - \sigma_3)/2$$

$$S_3 = B_2 e^{-\eta L_2} - (\sigma_2 - \sigma_3)/2$$

Figure 5.13 shows the variation of channel potential as a function of position along the channel length from source to drain for various values of V_{GS}. It can be observed that the increase in V_{GS} shifts the potential upwards and creates more band bending. This will result in increase in tunneling current at the source junction while also increasing the ambipolar current occurring at the drain junction. In order to decrease the unwanted ambipolar current and increase the tunneling current, gate work function engineering techniques have been incorporated.

The gate has been composed of different work function materials, and the impact of these materials is shown in Figure 5.14. In the first case, gate materials M_1 and M_2 have been considered of the same work function material, whereas M_3 is considered of comparatively low work function material. In the second case, M_1 is low work

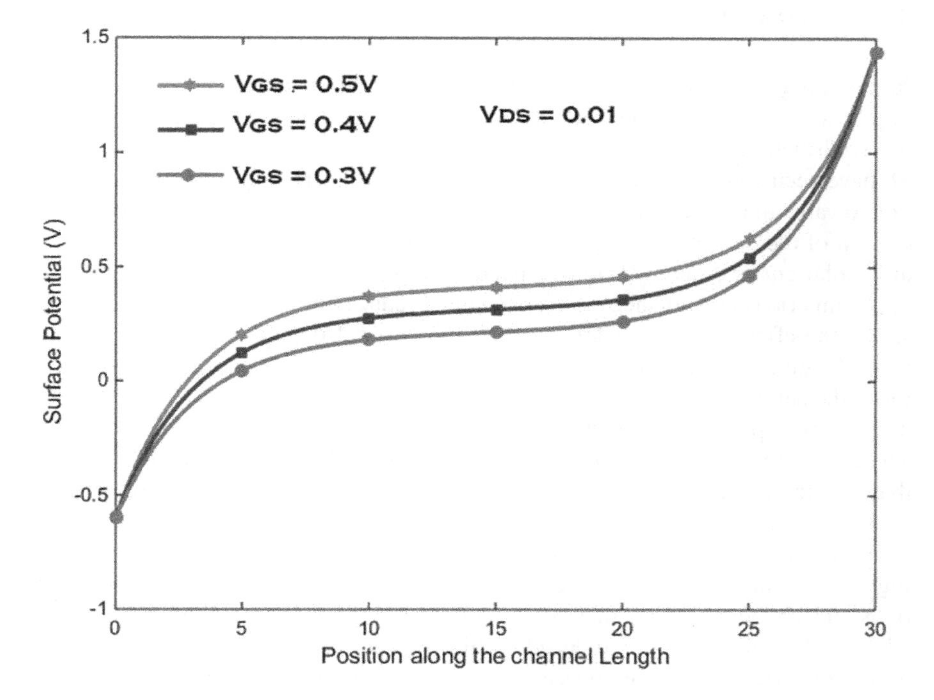

FIGURE 5.13 Analytical and simulated surface potential variation versus position along the channel length for different gate voltage.

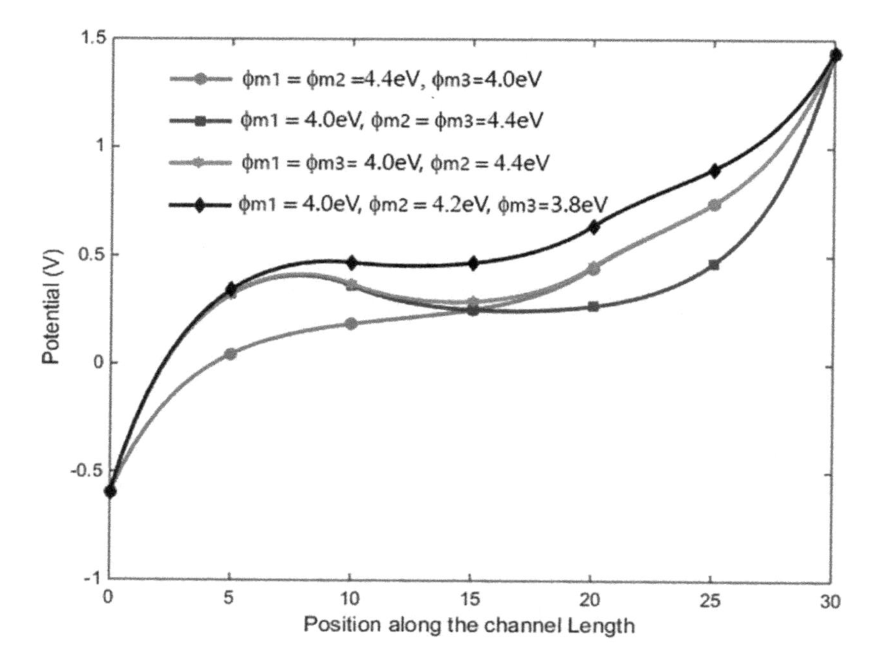

FIGURE 5.14 Analytical and simulated surface potential variation versus position along the channel length for different gate work functions.

function material, whereas M_2 and M_3 are of the same material with comparatively higher wok function. In the third case, M_1 and M_3 are considered low work function materials, whereas M_2 is high work function material. In the fourth case, M_1, M_2, and M_3 have been considered as different work function materials. The higher work function towards the source end reduces the band bending, and it will result in lower drive current of the device. The higher work function towards the drain end will increase ambipolar current due to increased band bending. Third and fourth cases represent optimum criteria to enhance drive current and reduce ambipolar current. Figure 5.15 shows the effect of trapezoidal shape on the channel potential. It can be observed from the given figure that the increase in the inclination angles of the side gate will lower the band banding and hence degrade the device performance. The case of inclination = 0^0 represents the rectangular device and shows maximum band bending. The effect of increase of V_{ds} on the device is shown in Figure 5.16, and it can be seen that the increase in V_{ds} has negligible effect on the source/channel tunneling junction.

The electric field variations in the channel region are shown in Figures 5.17–5.19. Figure 5.17 shows the electric variation for increasing values of V_{GS}, whereas Figure 5.18 shows the electric field variation for different gate material work functions. It can be observed that M_1 is composed of low work function materials, whereas M_2 and M_3 are composed of high work function material, which shows higher electric field at the source/channel tunneling junction while reduced electric field at the drain/channel junction. The electric field can further be reduced by placing lower work-function gate material near the drain end to further reduce

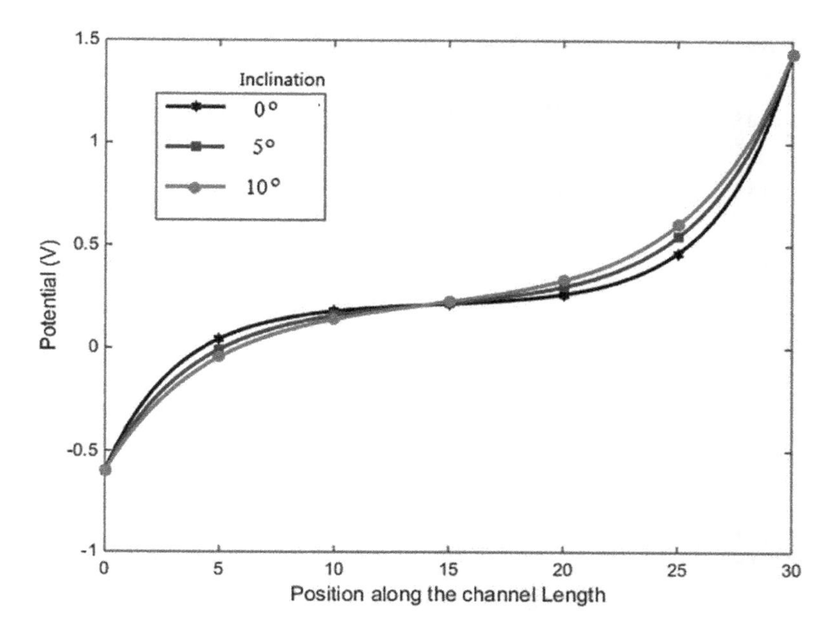

FIGURE 5.15 Analytical and simulated surface potential variation versus position along the channel length for different device inclinations.

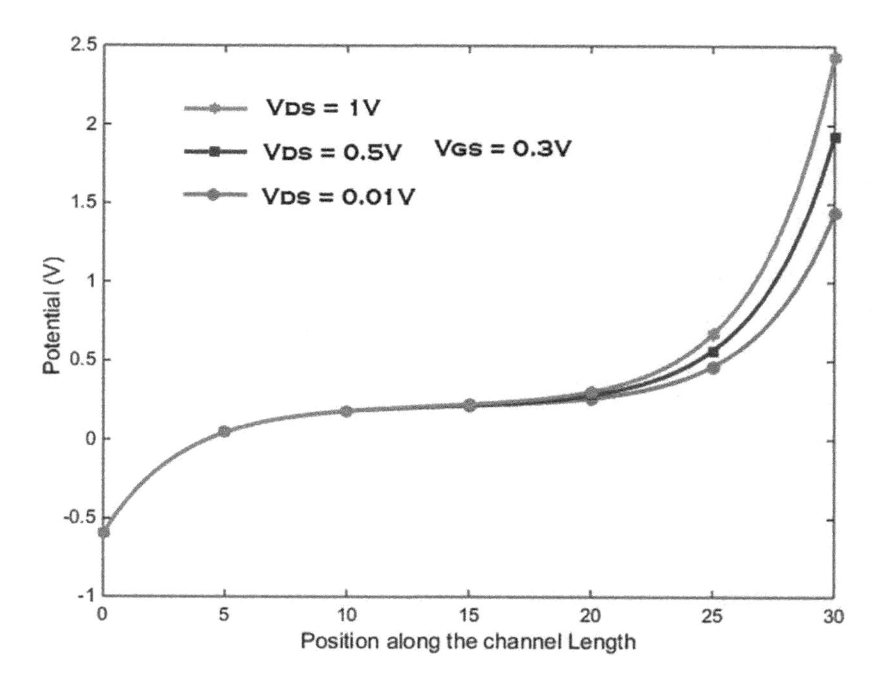

FIGURE 5.16 Analytical and simulated surface potential variation versus position along the channel length for different drain–source bias.

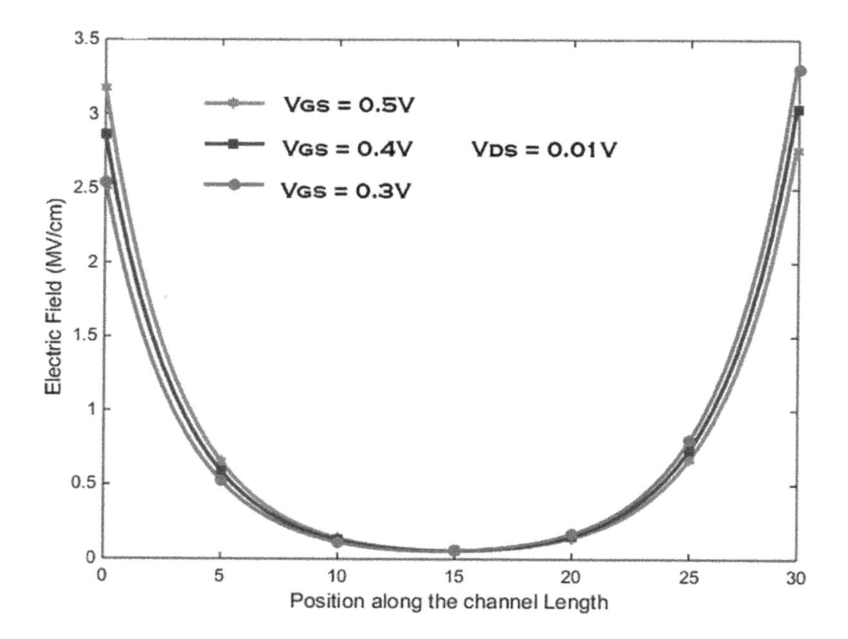

FIGURE 5.17 Analytical and simulated surface potential variation versus position along the channel length for different values of V_{GS}.

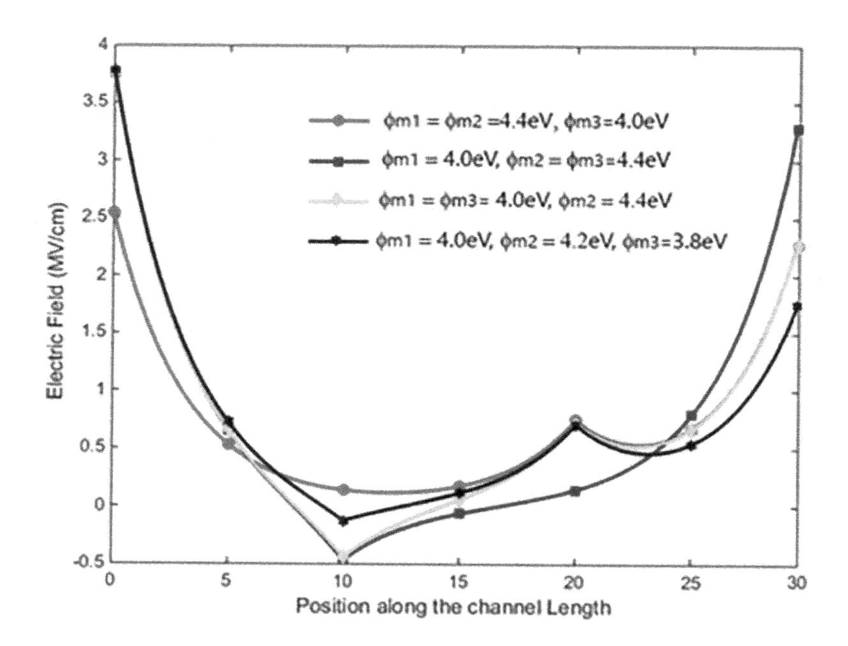

FIGURE 5.18 Analytical and simulated surface potential variation versus position along the channel length for different gate work functions.

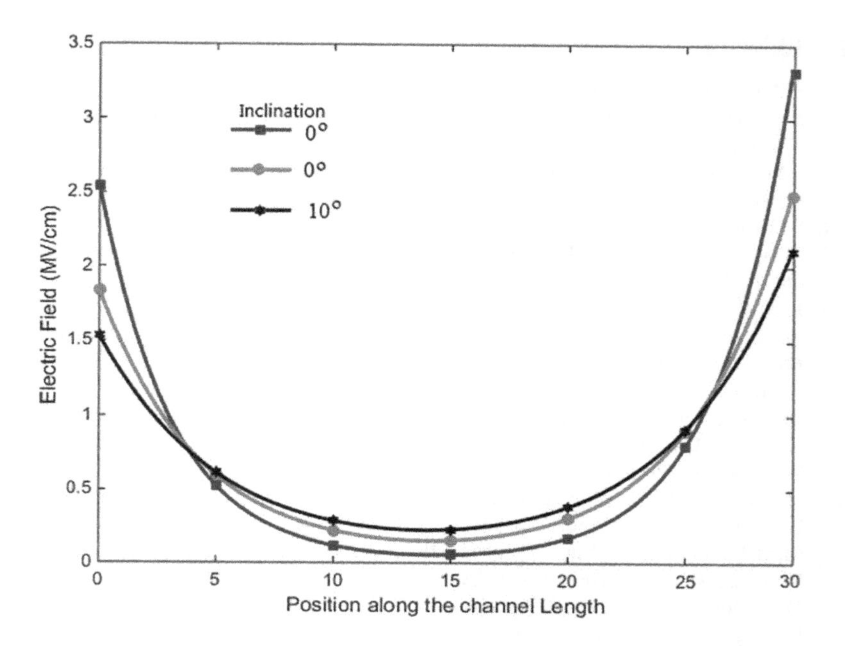

FIGURE 5.19 Analytical and simulated surface potential variation versus position along the channel length for different device inclinations.

the ambipolar current and other parasitic effects. Figure 5.19 shows the effect of inclination angle on the channel potential. The increase in inclination angles result in decrease in electric field near source/drain and channel junctions. The will reduce the current drive of the device. The use of gate work-function engineering can help to compensate for the low drive as already discussed.

5.6 SUMMARY

In nanoscale MOSFET, several short channel and electrostatic effects come into play that make the conventional MOSFET non-functional. Therefore, to extend the Moore's law in nanometer range, several techniques/structure have been given to mitigate these effects. Among these structures, multigate MOSFETs dominate these structures to extend the ITRS predictions in sub-100 nm range. The multigate structures include the double-gate, trigate, gate-all-around, and FinFET structures. The multigate MOSFET structures have been further aided by gate and channel engineering techniques to make MOSFET work below 50 nm range. The gate engineering techniques include the division of gate into multiple regions and the employment of metals with different work functions. The metal gates with different work functions are suitably placed to achieve better performance. The channel engineering techniques include the division of channel into multiple regions with different doping concentrations. The doping in the channel regions is suitably done to achieve better performance. The performance is further improved by taking the non-conventional MOSFET structures such as trapezoidal FET or tunnel FET.

REFERENCES

1. SIA roadmap (http://www.semiconductors.org/).
2. Y. Tsividis, *Operation and Modeling of MOS the Transistor*, McGraw Hill ISBN: 9780195170153, 1–752 (New York, NY, 1988).
3. S.M. Sze, K.K. Ng, *Physics of Semiconductor Devices*, 3rd ed., John Wiley & Sons ISBN8126517026, 9788126517022, 1–826 (Hoboken, 2007).
4. H. Brooks, *Adv. Electron. Electron Phys.* 7, 85–182 (1955).
5. M.A. Kharadi, G.F.A. Malik, K.A. Shah, F.A. Khanday, *IEEE Trans. Electron Devices* 66(11), 4976–4981 (2019).
6. J. Bardeen, W.H. Brattain, *Phys. Rev.* 74, 230–231 (1948).
7. W. Shockley, *Electrons and Holes in Semiconductor With Application to Transistor Electronics*", D. Van Nostrand ISBN 10: 0442075936 ISBN 13: 9780442075934 (New York, 1950).
8. J.E. Lilienfeld, F. Brooklyn, "Method and apparatus for controlling electric currents," US Patent US1745175A, New York, NY, USA, 1930.
9. W. Shockley, *Proc. IRE* 40, 1365–1376 (1952).
10. H.K.J. Ihantola, "Design theory of a surface field-effect transistor," Stanford Electronics Laboratories Tech. Report No. 1661-1, Stanford University, CA, 1961.
11. M.M.J. Atalla, D. Kahng, "Electric field-controlled semiconductor device," US Patent US3102230A, New York, NY, USA, 1963.
12. G.E. Moore, *Electronics* 38(8), 114 (1965).
13. F. Faggin, M.E. Hoff, S. Mazor, M. Shima, *IEEE Micro*, 16(6), 10–20 (1996).
14. R.H. Dennard, *J. Vac. Sci. Technol.* 19, 537–539, (1981).
15. J.R. Brews, W. Fichtner, E.H. Nicollian, S.M. Sze, *IEEE Trans. Electron Device Lett.* EDL-I, 2–4, (1980).
16. G. Baccarani. M.R. Wordeman, R.H. Dennard, *IEEE Trans. Electron Devices* 31, 452462 (1984).
17. C.G. Sodini, P.K. Ko, J.L. Moll, *IEEE Trans. Electron Devices* 31(10), 1386–1393 (1984).
18. G. Taylor, *IEEE Trans. Electron Devices* 30(8), 871–876 (1983).
19. R.R. Troutman, *IEEE Trans. Electron Devices* ED-26(4), 461–468 (1979).
20. Y. Taur, T.H. Ning, *Fundamentals of Modern VLSI Devices*, 2nd ed., Cambridge University Press, ISBN: 110739399X, 9781107393998, 1–1156 (Cambridge, 2009).
21. J.G. Ruch, *IEEE Trans. Electron Devices* 19(5), 652–654 (1972).
22. N.D. Arora, M.S. Sharma, *IEEE Trans. Electron Devices Lett.* 13(2), 92–94 (1992).
23. H. Wong, H. Iwai, *Microelectron. Eng.* 83, 1867–1904 (2006).
24. N. Kotani, S. Kawazaki, *Solid-State Electron.* 22(1), 63–70 (1979).
25. T.N. Nguyen, J.D. Plummer, *IEDM Technical Digest*, 596–599 (1981).
26. N.R. Santiago "Characterization, Modelling and Simulation of Deca-nanometer SOI MOSFETs", Ph.D. thesis, University of Granda, 2007.
27. S. Veeraraghavan, J.G. Fossum, *IEEE Trans. Electron Devices* 36(3), 522–528 (1989).
28. L.T. Su et al., *IEEE Trans. Electron. Devices* 41(1), 69–75 (1994).
29. T. Ernst, D. Munteanu, S. Cristoloveanu, T. Ouisse, N. Hefyene, S. Horigushi, Y. Ono, Y. Takahashi, K. Murase, "Ultimately thin SOI MOSFETs: Special characteristics and mechanisms", *Proceedings IEEE International SOI Conference*, pp. 92–93, 1999.
30. E. Suzuki, K. Ishii, S. Kanemaru, T. Maeda, T. Tsutsumi, T. Sekigawa, K. Nagai, H. Hiroshima, *IEEE Trans. Electron Devices* 47, 354–358, (2000).
31. F. Gámiz et al., *Solid-State Electron.* 45(4), 613–620 (2001).
32. C. Hu, "SOI and nanoscale MOSFETs", *Device Research Conference*, IEEE, pp. 3–4, 2001.
33. J. Colinge, *Solid-State Electrons* 48, 897–905 (2004).

34. I. Ferain, C.A. Colinge, J.-P. Colinge, *Nature* 479, 310–316, 2011. DOI: 10.1038/nature10676.
35. T. Sekigawa, Y. Hayashi, *Solid-State Electron.* 27(8–9), 827–828 (1984).
36. F. Balestra et al., *IEEE Electron Device Lett.* 8(9), pp. 410–412 (1987).
37. A.T. Shora, F.A. Khanday, *Int. J. Electron. Lett.*, DOI: 10.1080/21681724.2019.1600729.
38. D. Hisamoto, W.-C. Lee, J. Kedzierski, H. Takeuchi, K. Asano, C. Kuo, E. Anderson, T.-J. King, J. Bokor, C. Hu, *IEEE Trans. Electron Devices* 47(12), 2320–2325 (2000).
39. B. Doyle, B. Boyanov, S. Datta, M. Doczy, S. Hareland, B. Jin, J. Kavalieros, T. Linton, R. Rios, R. Chau, *VLSI Technol. Symp. Tech. Dig.*, 133–134 (2003).
40. J.P. Colinge, *FinFETs and Other Multi-Gate Transistors*, Springer Science + Business Media (Berlin, 2007).
41. A.T. Shora, F.A. Khanday, *J. Semicond.* 39(12), 1–6 (2018).
42. A.T. Shora, F.A. Khanday, *IET Circuits Devices Syst.* 13(8), 1107–1116 (2019). DOI: 10.1049/iet-cds.2018.5302D.
43. D. Jimenez, J.J. Saenz, B. Iniguez, J. Sune, L.F. Marsal, J. Pallares, *IEEE Electron Device Lett.* 25, 314–316 (2004).
44. K. Akarvardar, S. Cristoloveanu, P. Gentil, B.J. Blalock, B. Dufrene, M.M. Mojarradi, "Depletion-all-around in SOI G4-FETs: A conduction mechanism with high performance", *Proceedings 34th ESSDERC*, pp. 217–220, 2004.
45. M. Cheralathan, "Compact modeling for multi-gate MOSFETs using advanced transport models", PhD. Thesis, 2013.
46. J.-P. Colinge, J.C. Alderman, W. Xiong, C.R. Cleavelin, *IEEE Trans. Electron Devices* 53(5), 1131–1136 (2006).
47. I.C. Chen, S.E. Holland, C. Hu, *IEEE Trans. Electron Devices* 32(2), 413–422 (1985).
48. J. Robertson, *J. Vac. Sci. Technol. B Microelectron. Nanometer Struct. Process. Meas. Phenom.* 18, 1785–1791 (2000).
49. B.H. Lee, J. Oh, H.H. Tseng, R. Jammy, H. Huff, *Mater. Today* 9(6), 32–40 (2006).
50. K.P. Pradhan, S.K. Mohapatra, P.K. Sahu, D.K. Behera, *Microelectron. J.* 45, 144–151 (2014).
51. A. Chaudhry, M.J. Kumar, *IEEE Trans. Device Mater. Reliab.* 4(1), 99–109 (2004).
52. C.C. Hobbs et al., *IEEE Trans. Electron Devices* 5(6), 971–977 (2004).
53. R. Rios, N.D. Arora, C-L. Huang, *IEEE Trans. Electron Devices* 42, 935–943 (1995).
54. Y.C. Yeo, P. Ranade, T.J. King, C. Hu, *IEEE Electron Device Lett.* 23(6), 342–344 (2002).
55. G. Lucovsky, *Microeletron. Reliab.* 43, 1417–1426 (2003).
56. R. Chau, S. Datta, M. Doczy, B. Doyle, J. Kavalieros, M. Metz, *IEEE Electron Device Lett.* 25(6), 408–410 (2004).
57. W. Long, H. Ou, J. Kuo, K.K. Chin, *IEEE Trans. Electron Devices* 46(5), 865–870 (1999).
58. X. Zhou, *IEEE Trans. Electron Devices* 47(1), 113–120 (2000).
59. T.K. Chiang, M.L. Chen, *Japn. J. Appl. Phys.* 46(6A), 3283–3290 (2007).
60. S. Naskar, S. Sarkar, *IEEE Trans. Electron Devices* 60(9), 734–2740 (2013).
61. P. Razavi, A. Orouji, "Nanoscale triple material double gate (TM-DG) MOSFET for improving short channel effects", *International Conference on Advances in Electronics and Microelectronics*, pp. 11–14, 2008.
62. S. Deb, N. Singh, N. Islam, S. Sarkar, *IEEE Trans. Nanotechnol.* 11(3), 472–478 (2012).
63. P. Banerjee, S. Sarkar, *IEEE Trans. Electron Devices* 64(2), 368–375 (2017).
64. B. Cheng, A. Inani, V.R. Rao, J.C.S. Woo, "Channel Engineering for High Speed Sub-1.0 V Power Supply Deep Sub-Micron CMOS", *Technical Digest, Symposium on VLSI Technology*, pp. 69–70, 1999.
65. Q. Xie, C.J. Lee, J. Xu, C. Wann, J.Y.C. Sun, Y. Taur, *IEEE Trans. Electron Devices* 60(6), 1814–1819 (2013).

66. Y. Sun, S.E. Thompson, T. Nishida, *AIP J. Appl. Phys.* 101(10), 104503 (2007).
67. E. Ungersböck, S. Dhar, G. Karlowatz, V. Sverdlov, H. Kosina, S. Selberherr, *IEEE Trans. Electron Devices* 54, 2183 (2007).
68. S. Datta et al., "High mobility Si/SiGe strained channel MOS transistors with HfO_2/TiN gate stack", *IEEE International Conference on Electron Devices Meeting*, pp. 28.1.1–28.1.4, 2003.
69. A.T. Shora, F.A. Khanday, "Analytical Modelling for nanoscale Gate Engineered Silicon-On-Nothing MOSFET with High-K dielectric", *3rd International Conference on Communication and Electronics Systems (ICCES 2018)*, 15–16 October 2018, PPG Institute of Technology, Coimbatore, India. IEEE Xplore Part Number: CFP18AWO-ART; ISBN: 978-1-5386-4765-3, pp. 212–216, 2018.
70. Atlas User's Manual, *SILVACO Int.*, Santa Clara, CA, USA, 2016.
71. W.Y. Choi, B.G. Park, J.D. Lee, T.J.K. Liu, *IEEE Trans. Electron Devices* 28(8), 743–745 (2007).
72. M.G. Bardon, H.P. Neves, R. Puers, C.V. Hoof, *IEEE Trans. Electron Devices* 57(4), 827–834 (2010).
73. J. Cao, Ph.D. dissertation, Université Grenoble Alpes, France, 2017.
74. P. Saha, S. Sarkhel, P. Banerjee, S.K. Sarkar, "3D modeling based performance analysis of gate engineered Trigate SON TFET with SiO_2/HfO_2 stacked gate oxide", *IEEE International Conference on Electronics, Computing and Communication Technologies (CONECCT)*, Bangalore, India, 2018.
75. W.G. Vandenberghe, A.S. Verhulst, G. Groeseneken, B. Soree, W. Magnus, "Analytical model for a tunnel field-effect transistor", *Proceedings of IEEE MELECON Conference*, Ajaccio, France, pp. 923–928, 2008.
76. S. Marjani, S.E. Hosseini, R. Faez, *J. Comput. Electron.* 15(3), 820–830 (2016).
77. K.K. Young, *IEEE Trans. Electron Devices* 36(2), 399–402 (1989).
78. A.T. Shora, F.A. Khanday, *Int. J. Numer Model. Electron. Networks Devices Fields* 32(03), e2571 (2019). DOI: 10.1002/jnm.2571.

6 Spin Nanoscale Electronic Devices and Their Applications

6.1 INTRODUCTION TO SPINTRONICS

In any material, the conduction of current is considered due to free carriers. Conventionally, the charge of free carrier is used to describe the amount of current. In order to achieve high functionality of semiconductor chips, scaling of devices has been the main tool. The high-density chips require the low power operation as well. Therefore, several techniques and charge-based devices were reported to achieve this goal. In pursuit of achieving low-power devices with better or more functionality, the researchers started to consider the spin property of carriers instead of charge, which was earlier considered to play fundamental role in magnetism [1]. While magnetism remains the domain of spin, it was realized in the late 20th century that spin, alone or in conjunction with charge, can be harnessed to process information, particularly digital information encoded with binary bits 0 and 1 [1]. This lead to the emergence of a new field of electronics called "spintronics." For any operation in spintronics, the techniques for injection, detection, manipulation, transport, and storage of spins need to be established. In conventional solid-state charge-based devices, the size of the devices is scaled and atomic level proves to be the ultimate limit of scaling where the quantum effects dominate. In spin-based devices, encoding and reading out spin information in single spins can be considered the ultimate limit for scaling magnetic information [2]. Spintronics promises to be a technology of beyond-CMOS computing. Various spin devices such as non-volatile magnetic random access memory (MRAM) [3], programmable spintronic logic devices based on magnetic tunnel junction (MJT) elements [4], rotational speed control systems [5], and positioning control devices in robotics have been proposed and studied for various applications [6]. These developments have been possible due to the discovery of important phenomena of giant magnetoresistance (GMR) [7] and tunnel magnetoresistance (TMR) [8,9]. Spin field-effect transistor (spin-FET) is another class of spin-based device which is believed to be a better device than conventional semiconductor field effect transistor devices due to its exceptional properties like control of conductivity using spin degree of freedom and can offer high degree of integration, low power dissipation, high value of transconductance (for high-speed applications), the capability of high amplification (voltage, current, and power gains), etc. [1]. Owing to less energy required for control of spin, spin-FET is well suited for low-power applications and hence can resolve the power issues of conventional transistors [1]. The spin-based computing in semiconductors offers a pathway towards integration of information storage and processing in a single material. In other words, logic-in memory computing is possible in spintronics, which lead to their use in bio-inspired applications [10,11].

6.1.1 Giant Magnetoresistance (GMR) and Its Applications

The "giant magnetoresistance" (GMR) effect was discovered in the late 1980s by two European scientists working independently: Peter Gruenberg of the KFA Research Institute in Julich, Germany, and Albert Fert of the University of Paris-Sud. They saw very large resistance changes 6% and 50%, respectively, in alternating very thin layers of various metallic elements. They performed experiments at low temperatures and in the presence of very high magnetic fields. The materials used were laboriously grown that cannot be mass-produced. Given the scope of applications of GMR effect, scientists around the world started the research for further development of this effect and possible applications [12,13].

The electric current in a sequence of thin ferromagnetic (FM) layers separated by equally thin non-magnetic (NM) metallic layers, as shown in Figure 6.1, is strongly influenced by the relative orientation of the magnetizations of the magnetic layers [9]. The resistance of the magnetic multilayer is low when the magnetizations of all the magnetic layers are parallel, but it becomes much higher when the magnetizations of the neighboring magnetic layers are ordered antiparallel. This suggested that the internal magnetic moment of electrons associated with their spin plays an important role in transport of electric charge.

GMR effect is based on spin-dependent scattering in specific FM/NM/FM structure, and the scattering probability depends on the magnetization direction of the FM layers. The basic mechanism of GMR effect was explained by Mott's two-channel model [14]. The electrons can be divided in two channels: spin-up electron channel and spin-down electron channel. The schematic representation of this model is shown in Figure 6.2.

- When the two FM layers have their relative magnetization orientations in parallel, the spin-up electrons experience less scattering and pass through the three-layer structure with least resistance R_L, whereas the spin-down electrons experience significant scattering in both FM layers and hence see high resistance R_H. The total resistance in this case is given by

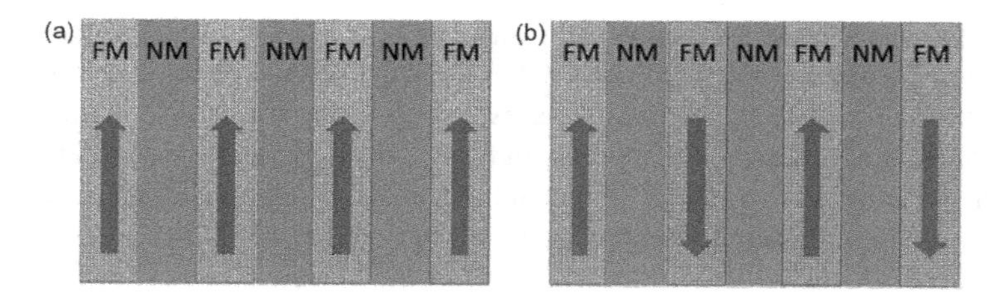

FIGURE 6.1 (a) Parallel FM configurations of magnetic multilayer film. (b) Antiparallel FM configurations of magnetic multilayer film.

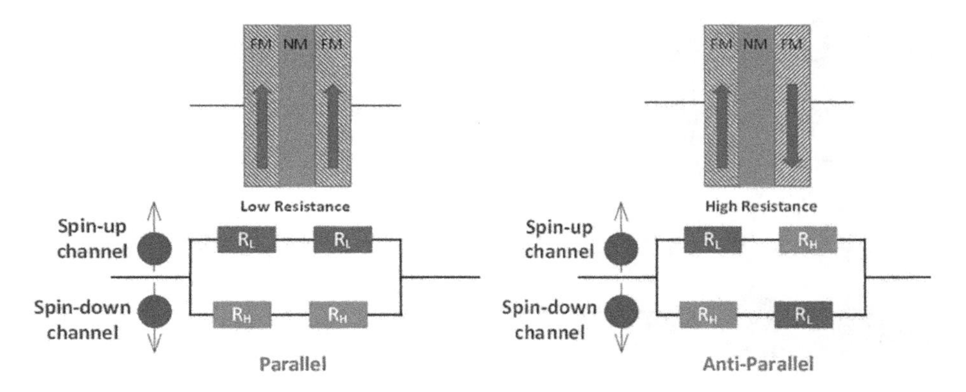

FIGURE 6.2 Schematic of Mott's two-channel model.

$$R_P = 2R_H R_L / (R_H + R_L)$$

- When the relative magnetization direction of two FM layers is antiparallel, both spin-up and spin-down electrons suffer from scattering in either FM layer resulting in high resistance given by

$$R_{AP} = (R_H + R_L)/2$$

The GMR ratio is defined by

$$\text{GMR} = \Delta R/R_P = (R_H - R_L)^2 / 4R_H R_L$$

The other types of GMR include spin-valve GMR, pseudo-spin GMR, and granular GMR [15].

The discovery of GMR effect did not only give birth to a new field, but also presented itself as a model that pushed the fundamental research towards industrial products. The first commercial GMR sensor was announced in 1994 [16]. Nowadays, GMR sensor is used in data storage, biological applications, space applications, etc. [16]. The first hard disk drive (HDD) with GMR read head produced by IBM in 1994 increased more than ten times in the storage density [17]. Currently, the storage density of HDD based on GMR effect is more than 500 Gb/in^2. GMR effect was also exploited in the development of MRAM until the discovery of tunneling magnetoresistance effect (TMR), which shows more advantages in MRAM applications.

6.1.2 Tunnel Magnetoresistance (TMR) and Its Applications

TMR was first observed by Jullière in Fe/Ge/Co junction in 1975 [9]. Conductance measurement depends on the spin polarizations of FM layers. As shown in Figure 6.3, a spintronic device which consists of two FM layers separated by a thin insulating barrier. The magnetization orientation of one of the layer is kept fixed by means of an anti-FM layer known as pinning layer, and the spin orientation of the other layer can

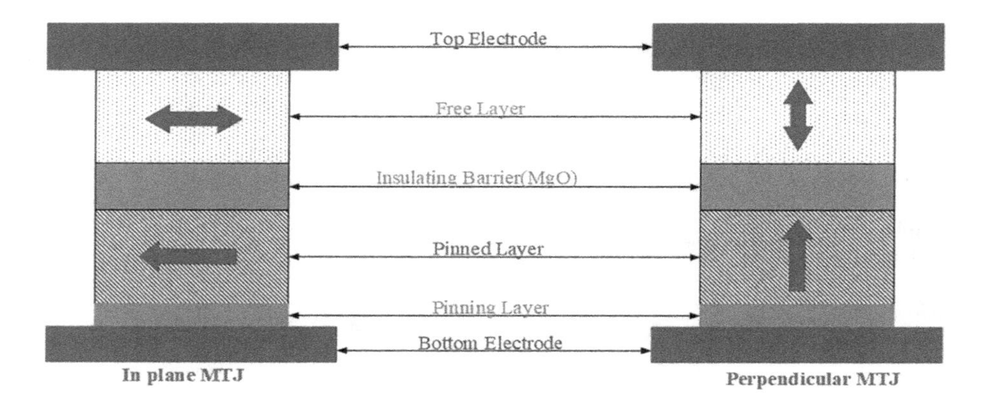

FIGURE 6.3 Schematic of MTJ structure.

be changed freely. The layer whose magnetization orientation is fixed is known as pinned layer (or fixed layer or reference layer) and that whose magnetization orientation can be changed is known as free layer. Such a device is called MTJ. The magnetization of the FM layers can be either in the film plane or perpendicular to the film plane, which defines two kinds of MTJs: in-plane MTJs and perpendicular MTJs.

Depending upon the relative magnetization orientations of free layer and pinned layer, MTJ exhibits two resistance states: parallel and anti-parallel. When the magnetization orientations of two layers are parallel, MTJ exhibits a low-resistance state denoted by R_P, and when the magnetization orientations are antiparallel, it exhibits a high resistance state denoted by R_{AP}. These two states are considered as two logic states where R_P represents logic "1" and R_{AP} represents logic "0." The fundamental difference between a GMR structure and MTJ structure is that the non-magnetic metallic layer in former is replaced in later by a thin insulating barrier of thickness of about 1 nm. The fractional change in resistance of MTJ from antiparallel to parallel state is defined as TMR given by

$$\mathrm{TMR} = \left(R_{AP} - R_P\right)/R_P$$

A high TMR ratio is a desired parameter for both logic as well as memory applications so as to get appreciable distinction between parallel and antiparallel states of MTJ. With MgO barrier, TMR can reach up to 500%. Most practical MTJs have TMR ratio between 50% and 150%.

6.1.3 Spin Injection Efficiency

The materials on the net spin polarization can be generally classified as paramagnets and ferromagnets, where paramagnets have equal population of up-spin and down-spin electrons under equilibrium, whereas the ferromagnets have non-zero spin polarization [1]. In spintronics, majority of the devices use both paramagnetic and FM materials. In paramagnetic materials used in these devices, the net spin polarization is mostly obtained by electrical injection of spin-polarized carriers from the FM materials. The efficiency of injecting only one type of carriers (spin-up or spin-down) from FM material into the paramagnetic material is measured by spin injection efficiency. The ideal value of spin injection efficiency is 100%; that is, only one type of carrier (spin-up or spin-down) is injected at the ferromagnet/paramagnet interface. However, in practice both spin-up and spin-down carriers are injected at the interface but not equally. Therefore, several works have been reported in the literature to enhance this efficiency [18–20]. Efficiencies up to 90% have also been demonstrated which even may not be sufficient for practical spin devices such as spin-FET [19].

Since both up-spin and down-spin carriers will be present in a paramagnetic material, a FM material should be able to filter out only one type of spin carrier from a paramagnetic material. This is measured by spin detection efficiency. Like the spin injection efficiency, the ideal value of spin detection efficiency is 100%; that is, only one type of spin carriers transmits from paramagnetic material into the FM material.

6.2 SPIN DEVICES

6.2.1 Magnetic Tunnel Junction (MTJ)

MTJ is one of the most promising spintronics devices and applied as the basic memory cell in MRAM and magnetic logic development. It is a non-volatile FM device, which is formed by sandwiching of two FM material layers with a fixed layer pinned to constant direction with other pointing bi-directional free layer/storage layer, as depicted in Figure 6.3. These FM layers are separated by thin non-magnetic insulating material. The insulating material is so thin that electrons can travel through it by tunneling. The spin polarization in the pinned layer and storage layer is in the same or opposite direction. The configuration of the MTJ stack can be switched simply by altering the spin magnetization direction of the storage layer, which may be induced by the magnetic field with opposite direction and superior than the threshold value. The phenomenon of change in resistance due to an externally applied field is explained by tunnel TMR effect. To support fast writing speed and high efficiency in terms of power, one of the promising switching techniques is spin transfer torque (STT) in which the orientation of magnetization can be modified using a spin polarized current [21]. This basic physical technique enables simplifying CMOS circuitry. It also presents the advantages of lower threshold current and hence low-power, higher-speed operation, scalability, and high endurance [4].

MTJs are very likely the most successful spintronics device so far, emerging as the preferred building block of spintronic circuits mostly due to their large TMR ratios (thus allowing a large read-out signal) and the possibility of integration with conventional semiconductor electronics [22]. The change of the resistance depending on the relative alignment of the free and fixed layers in MTJs is due to the TMR effect discussed earlier. If we consider a typical FM metallic material such as Fe or Co, one of the conduction bands will have a lower energy (the majority band in Figure 6.4a) and therefore the majority of the electrons will have their spins aligned into this preferred direction, translating into a total net spin polarization (magnetization) in the FM material. If the magnetizations of both FM electrodes in the device

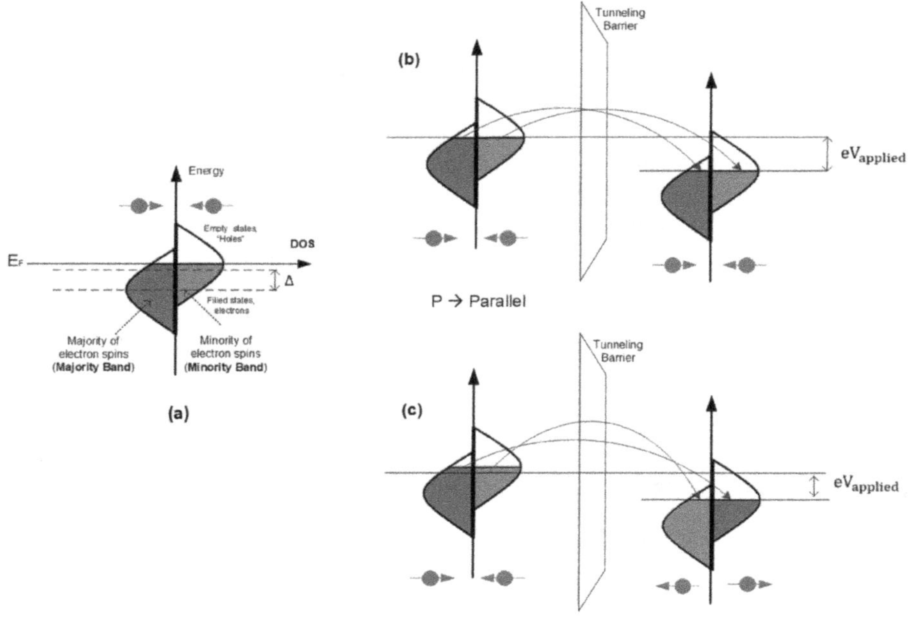

FIGURE 6.4 (a) Energy band diagram of FM material. (b) Energy band diagram when the magnetizations of both FM electrodes in the device are in the same direction. (c) Energy band diagram when the magnetizations of both FM electrodes in the device are in the opposite direction.

are in the same direction, the transport process is described in Figure 6.4b. In this scenario, a large number of electrons can tunnel to a large number of available states in one of the channels, resulting in a small resistance for this channel, and an overall small resistance since the transport will be dominated by the conduction through this path. Hence, the resistance of the MTJ will be small when the magnetizations are parallel to each other; otherwise, the overall resistance will be high as shown in Figure 6.4c. In conclusion, the dependence of the tunneling current on the relative magnetization directions of both free and fixed layers (FM electrodes) is responsible for the TMR observed in MTJs.

So far, the CoFeB|MgO|CoFeB "sandwich" in MTJs has developed into the equivalent to the Si|SiO_2 combination for the CMOS transistor community. This stack is most frequently used to build MTJ devices because of several practical advantages:

1. The stack can be fully sputter deposited, including depositions over large-scale wafers.
2. The MTJ is composed by backend of the line (BEOL) friendly materials, making it compatible with the fabrication of conventional semiconductor technology.

3. The usage of amorphous (as deposited) CoFeB allows the MgO to grow (001) oriented on top of it. Then, when the devices are annealed, the boron diffuses out and allows the CoFeB to crystallize bcc (001) next to the already oriented MgO, with lattice constants very close to each other [23]. This results in TMR ratios for MTJ devices with in-plane magnetization as high as 604% at room temperature. A thorough summary of the TMR ratios obtained for different in-plane and perpendicular MTJ systems is given in [24].

6.2.1.1 Switching Mechanism in MTJ

Knowing that MTJ can exhibit two resistance states R_P and R_{AP} based on TMR effect, it becomes important to know the methods by which the novel device can be switched between the states. There are few methods available, which are discussed below, to achieve the transition between the two states each having its own advantages and disadvantages.

6.2.1.1.1 Field Induced Magnetic Switching (FIMS)

Conventional MTJs use this type of switching method to switch the magnetization direction of free layer. In this method, the magnetic fields are generated by two orthogonal current carrying lines. Figure 6.5 illustrates the FIMS mechanism in MRAM. When an MTJ is selected to be switched, currents are produced in the orthogonal lines, which generate the magnetic fields. When the currents are large enough and the vector sum of the fields is sufficient, the magnetization orientation of the free layer in the selected MTJ is switched. However, this approach suffers from half-selectivity issue; that is, the MTJs (half selected) lying along the orthogonal current lines also experience some magnetic field and hence may have an opportunity of undesirable switching. Furthermore, the high currents (>10 mA) required to generate magnetic fields yield considerable power consumption and the electromigration

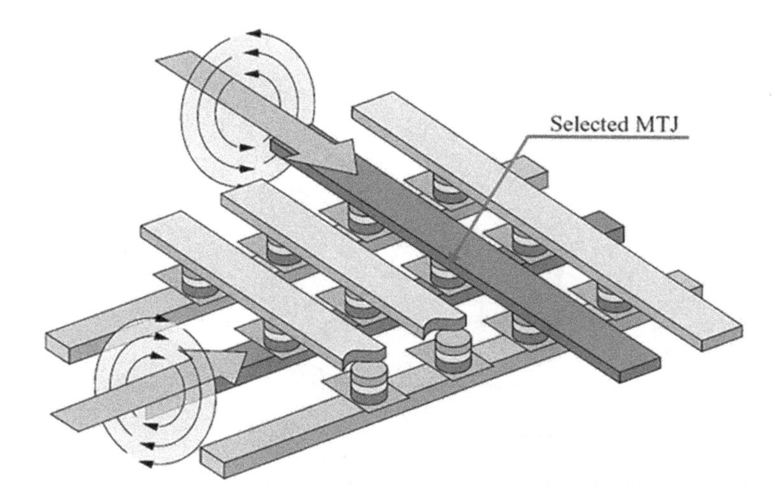

FIGURE 6.5 Field-induced magnetic switching in MRAM.

effect limits its scalability. These issues hindered its commercialization. Toggle switching proposed by Engel et al. from Freescale [25] could solve the half selectivity issue and based on this advanced switching method, MRAM was commercialized in 2006 [26] but the questions of speed, density and power consumption could not be answered by this method even.

6.2.1.1.2 Thermally Assisted Switching (TAS)

In order to address the issues faced by FIMS, thermally assisted switching (TAS) method was proposed [27]. Two extra anti-FM layers are added to the conventional MTJ, one (AF1) above the free layer and other (AF2) below the reference layer, with AF1 having higher blocking temperature than AF2. As shown in Figure 6.6, a temporary joule heat is produced by a pulse of low current I_h through the selected MTJ that heats the magnetic layers above their magnetic ordering temperature to reduce greatly the required switching field and then a magnetic field H, produced by magnetization current I_m, is applied to write it. This method of writing overcomes the half-selectivity problem and promises relatively lower power, higher density, and higher thermal stability as compared to FIMS approach. However, the mandatory heating and cooling processes lower the operation speed, which make TAS approach expensive in the high-speed logic applications.

6.2.1.1.3 Spin Transfer Torque (STT) Switching

The third switching methodology is based on STT, which uses only one low current going through the MTJ to switch its state. John Slonczewski at IBM first discovered this impact in 1996. STT switching is predicted as the most effective writing approach for MRAM and magnetic logic application till date. Figure 6.7 illustrates the STT switching approach, which demonstrates the two different states of MTJ. The state of the MTJ depends on the direction of current flow between the free and fixed layer. Writing current flowing from free layer to the fixed layer will store the logic "0" in the MTJ with the storage layer spin magnetization direction parallel to the pinned layer, whereas when current flows from the fixed layer to free layer, it will

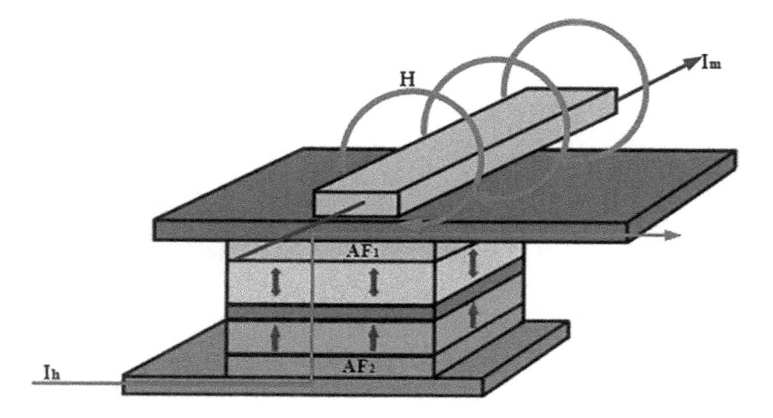

FIGURE 6.6 Thermal-assisted switching for MTJ.

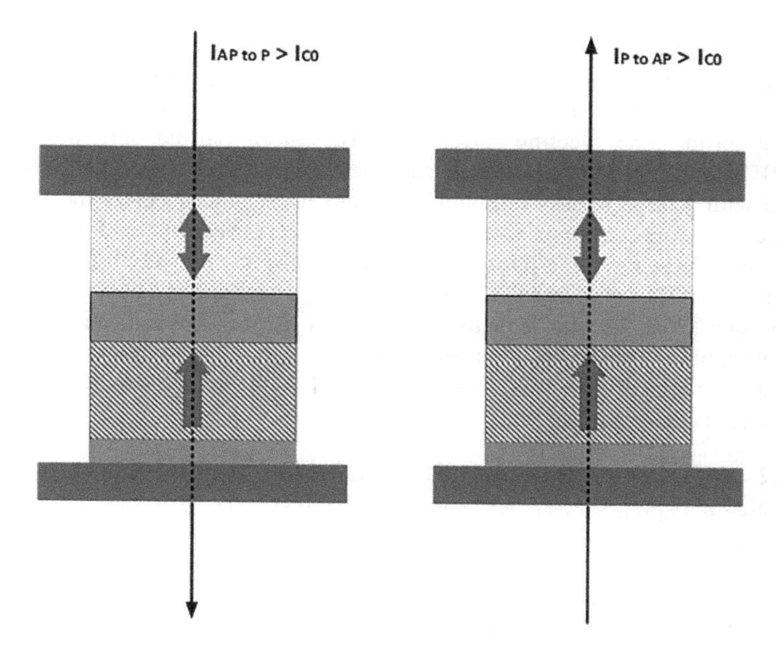

FIGURE 6.7 STT mechanism to change MTJ state.

store the logic "1" in the MTJ with the storage layer spin magnetization direction antiparallel to that of the pinned layer resulting in high resistance of the MTJ. In order to change the resistance state or logic-in MTJ, the writing current flowing through MTJ must be greater than the critical current density (J_{co}).

STT switching behavior can be categorized into two main regions:

1. Precessional region (IMTJ > IC)
2. Thermal activation region (IMTJ < IC).

In the precessional region, MTJ experiences a rapid precessional switching. In the thermal activation region, although the current is less than the critical value, the switching can occur with a long input current pulse due to the thermal activation. While STT approach offers significant advantages in terms of read energy and speed, a significant incubation delay due to the pre-switching oscillation [28,29] incurs high switching energy.

6.2.1.1.4 Spin Hall Effect (SHE) Switching

Even though current-induced STT writing mechanism exhibits many attractive features, it still has some disadvantages for MTJs to be embedded in logic circuits where speed is critical. STT needs long incubation delay (several nanoseconds) at the initial switching stage, due to random thermal fluctuations [30–32]. The low switching speed greatly limits its development for faster computing system. Besides, large bi-directional current passing though the MTJ nanopillar leads to larger writing circuit and higher risk of barrier breakdown. Since the read and write of the two-terminal

MTJ device share the same current path, read and write operations should be separated, and the read current should be small enough to avoid erroneous writing.

Spin Hall effect (SHE) is another way to switch the magnetization of the free layer by an in-plane injecting current [31]. Three-terminal magnetic device based on SHE has been proposed as shown in Figure 6.8, where a heavy metal strip (e.g., Ta, Pt) with a large spin-orbit coupling parameter is placed below the free layer. When a current (I_{SHE}) passes through the heavy metal, electrons with different spin directions are scattered in opposite directions. The spin-orbit coupling converts the charge current into perpendicular spin current (I_S), generating a torque called spin-orbit torque (SOT, or spin Hall torque) to assist magnetization reversal [30]. The orientation of the free layer is controlled by the direction of the injecting current.

A comparison of switching schemes of MTJ in terms of various performance parameters is given in Table 6.1.

6.2.1.2 MTJ Models, Design, and Simulation

To explore the applications of MTJ, it is important that the device models employed for the MTJ based logic design be discussed.

6.2.1.2.1 Magnetic Logic Devices

A 45-nm CMOS predictive technology models (PTM) [33] are used to integrate with MTJ models to create logic-in memory architectures. The macromodel of MTJ [34] provided by the University of Minnesota is used for the implementation of logic-in memory architecture and precessional voltage control magnetic anisotropy (VCMA) model is used to use MTJ as a probabilistic spin device [35].

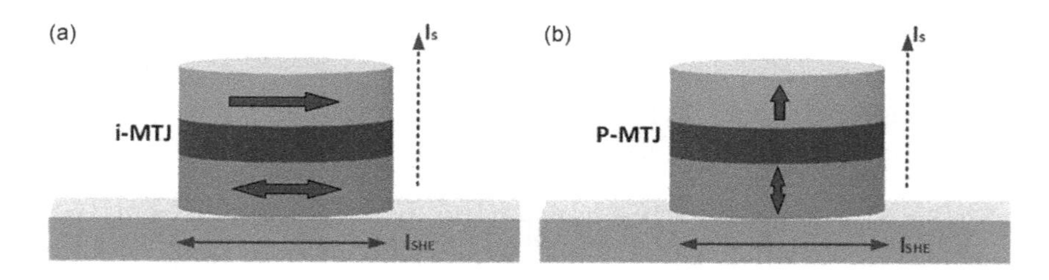

FIGURE 6.8 Three terminal magnetic device based on SHE using (a) I-MTJ and (b) P-MTJ.

TABLE 6.1
Comparison of Different Writing Approaches

Switching Technique	Switching Mechanism	Power	Speed	Area	Half-Selectivity	Read/Write Paths
FIMS	Field	High	Low	Large	Yes	Same
TAS	Field	Medium	Low	Medium	No	Same
STT	Current	Low	Medium	Small	No	Same
SHE	Current	Low	High	Small	No	Independent

There are two different types of parameters, which include predefined constants such as electron mass (m) and user-defined parameters such as the size of MTJ, which are listed in Tables 6.2 and 6.3, respectively.

TABLE 6.2
Predefined Parameters in MTJ SPICE Model

Parameter	Description	Value	Unit
E	Charge of electron	1.6×10^{-19}	C
M	Mass of electron	9.1×10^{-31}	kg
K_B	Boltzmann constant	1.38×10^{-23}	J/K
μ_B	Bohr magnetron constant	9.27×10^{-28}	J/Oe
$\hbar$	Reduced Plank's constant	1.0545×10^{-34}	JS
V_0	Fitting parameter	0.65	V
Γ	Gyromagnetic ratio	1.76×10^{7}	Hz/Oe
T_c	Curie temperature	1,420	K
B	Material-dependent constant	0.4	
α_{sp}	Material-dependent constant	2×10^{-5}	

TABLE 6.3
User-Defined Parameters in MTJ SPICE Model

Parameter	Description	Default Value	Unit
l_x	Width of free layer	65	nm
l_y	Length of free layer	130	nm
l_z	Thickness of free layer	1.8	nm
M_{s0}	Saturation magnetization at 0 K	1.075	Oe
P_0	Polarization factor at 0 K	0.715	
A	Gilbert damping factor	0.01	
Temp0	Ambient temperature	300	K
RA_0	Resistance area product	5.4	Ω μm^2
MA	Magnetic anisotropy	0 for IMTJ 1 for PMTJ	
Ini	Initial state	0 for P 1 for AP	
t_c	Critical thickness (only for PMTJ)	1.5 ($t_c > l_z$)	nm

6.2.1.2.2 Hysteresis

MTJs are two-terminal current-controlled hysteresis devices. The resistance of an MTJ changes when adequately high currents flow through the device. The MTJ can be in either the parallel or the antiparallel state. At the point when the current through the device surpasses the basic switching current, the state of the device can be switched. The discriminating switching currents for the two distinct states are not generally be indistinguishable, and it is simpler to change MTJ from the antiparallel to the parallel state [36]. The antiparallel state will have a higher cell resistance than the parallel state.

6.2.1.2.3 *Voltage Bias versus Resistance*

Figure 6.9a demonstrates that the device resistance drops with the increase in the applied voltage. This reliance can be approximated utilizing the Gaussian capacity [37]. The resistance was measured amid write pulse and incorporates the bias voltage effect.

6.2.1.2.4 *Critical Switching Current versus Critical Switching Time*

Critical switching currents are associated with critical switching time, and Figure 6.9b demonstrates the relationship between them for long pulses of writing time that are approximated by the following equation:

$$I_C = I_{C0}\left(1 - K_B/E\ln\left(\tau_p/\tau_o\right)\right) \tag{6.1}$$

where τ_o is set to 1 ns, and I_{C0} is the critical switching current plotted to 1 ns of pulse width. The thermal stability E/K_BT can be derived by differentiating I_C with respect to $\ln\left(\tau_p/\tau_o\right)$ and describes the device resistance to random switching due to thermal energy.

6.2.1.2.5 *Magnetic Dynamics of MTJ*

The Landau–Lifshitz–Gilbert (LLG) equation is a differential equation that describes the processional motion of a time-varying magnetization vector: $\bar{M} = \bar{M}(t) = \left[M_x(t),\ M_y(t),\ M_z(t)\right]$ and is given by

$$\frac{1+\alpha^2}{\gamma}\cdot\frac{d\bar{M}}{dt} = -\bar{M}\times\bar{H}_{\text{Keff}} - \alpha\cdot\bar{M}\times\left(\bar{M}\times\bar{H}_{\text{Keff}}\right) + \frac{\hbar PJ}{2et_F M_s}\cdot\bar{M}\times\left(\bar{M}\times\bar{M}_p\right) \tag{6.2}$$

where γ is the gyromagnetic ratio, α is the damping constant, $\hbar$ is the reduced Plank's constant, P is the spin polarization factor, e is the charge of electron, t_F is the thickness of free layer, J is the current density, M_s is the saturation magnetization, $\bar{M}_p$ is the magnetization vector of pinned layer, and $\bar{H}_{\text{Keff}}$ is the effective anisotropy field.

When there is no applied field, $H_{\text{app}} = 0$, the magnetization characteristics of the MTJ model are shown in the Figure 6.10.

6.2.1.3 Logic-In Memory Architecture

The scaling down of integrated circuits based on CMOS technology faces significant challenges due to technology advancing factors. The stand-by power becomes comparable to active power due to the increasing leakage current. Therefore, power gating and various other schemes are used to reduce the stand-by power. Another challenge is interconnection delay, as we know the von Neumann architecture separates logic and memory unit, thus needing a long interconnect, signal bottleneck,

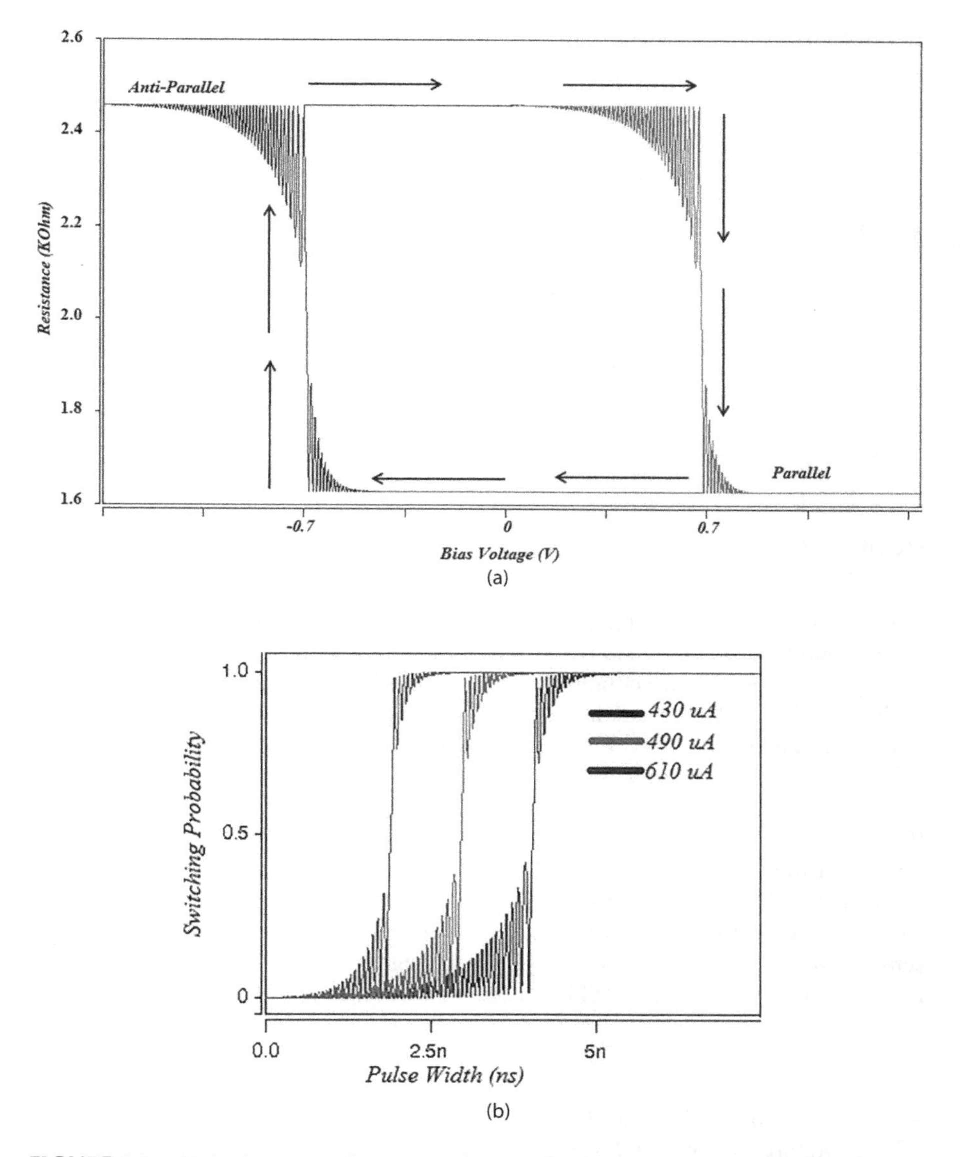

FIGURE 6.9 (a) Resistance–voltage curve for an MTJ. The resistance is low/high when the polarization of the magnetic layers is parallel/antiparallel. (b) Relationship between critical switching current magnitudes with critical switching time.

and power to charge and discharge those interconnects between these units within a processor. Most of the primary memory units are volatile which therefore needs constant voltage to store the information even in the presence of leakage currents. This signifies the need of logic-in memory architecture where the logic circuit is built based on the memory device. MTJ device has a feature of non-volatility, endurance, and high density, and is compatible with CMOS, which makes the implementation

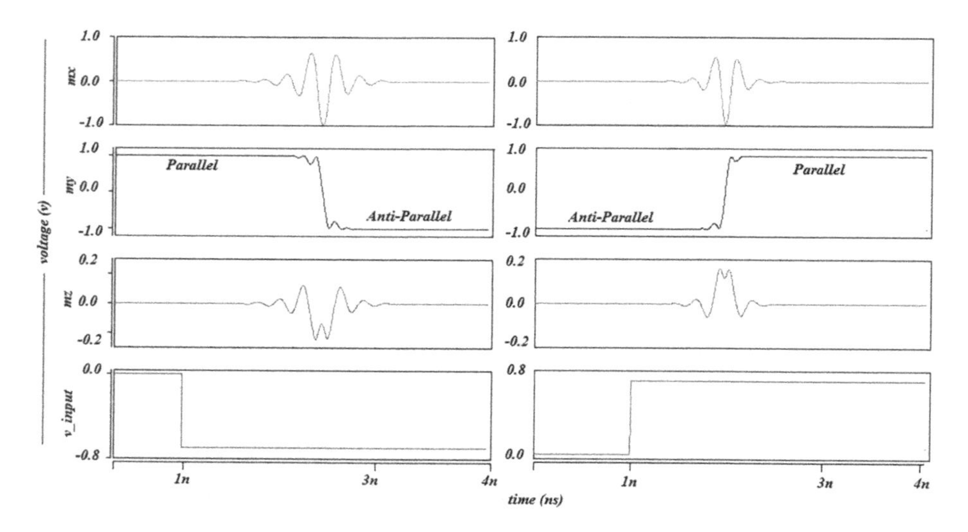

FIGURE 6.10 Components of magnetization vector and the voltage across the MTJ for switching from one region to another.

of next-generation logic-in memory chips possible. With all these features, it is possible to build hybrid architecture of CMOS/MTJ logic circuits and memories where the logic and memory are combined such that it brings non-volatility of MTJ to the design by which off-state leakage currents can be avoided. With recent developments in research and technology of integrated circuits, various high-speed arithmetic circuits with regular structures have been proposed.

6.2.1.3.1 Basic Structure of Logic-in Memory Architecture

To design the hybrid MTJ/CMOS logic circuit, logic-in memory architecture has been used [38]. MTJ is not suitable for direct logical output with integrated CMOS since it requires a sense amplifier to detect the state in it, so we use a pre-charged sense amplifier (PCSA) [39]. The basic block diagram of hybrid MTJ/CMOS circuit is shown in Figure 6.11. The MTJ/CMOS block consists of PCSA, NMOS transistor logic (volatile logic) integrated with non-volatile MTJ logic.

6.1.1.3.1.1 Pre-charged Sense Amplifier (PCSA) MTJ device presents the resistive property compatible with CMOS transistors, which enables the sensing of the MTJ configurations with CMOS amplifier circuits. Figure 6.12 presents a seven-transistor-based PCSA integrated with a two-MTJ complementary structure used in logic-in memory architecture, which operates in low power, high speed, and a low surface state. It consists of four P-MOS transistors (M_{P0} to M_{P3}), three N-MOS transistors (M_{N0} to M_{N2}) and two complementary MTJs connected with the configuration shown in Figure 6.12. The two MTJs are configured in a complementary mode to present one logic-bit; if one MTJ has antiparallel or high resistance, then the other MTJ will be a parallel or low-resistance state.

This sense amplifier works in two phases: *pre-charge* and *evaluate*. First, the complementary MTJs are pre-charged with logic "1" using a switching circuit when

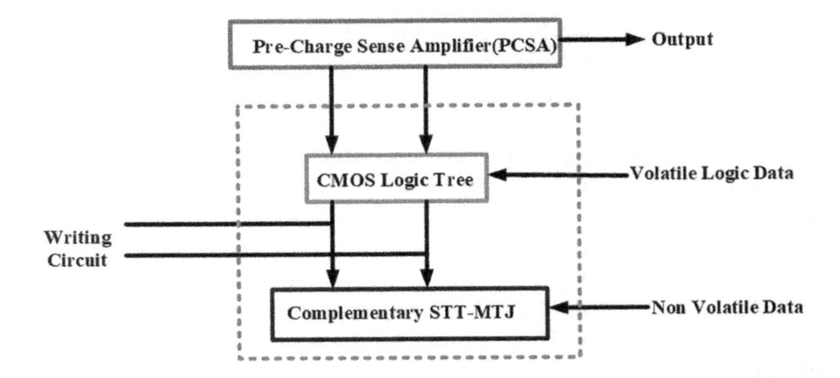

FIGURE 6.11 Block diagram of hybrid MTJ/CMOS logic-in memory architecture [38].

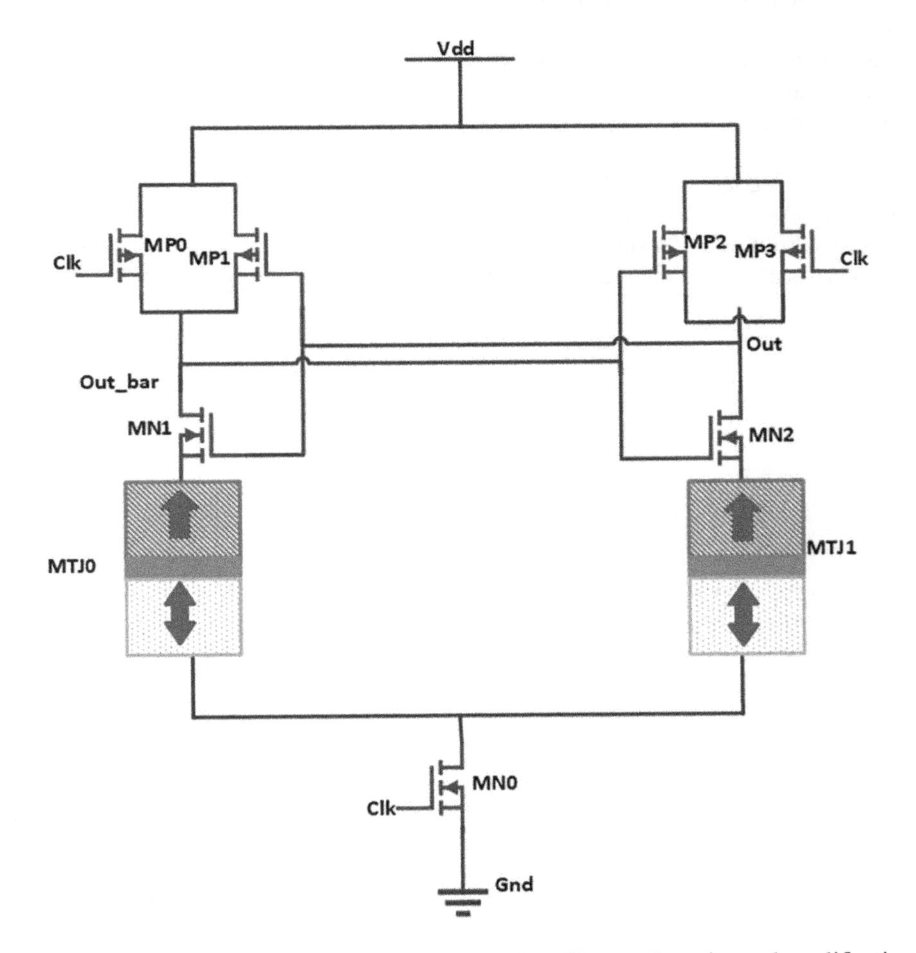

FIGURE 6.12 Pre-charge sense amplifier (PCSA) for MTJ state detection and amplification to logic level [38].

the clock input of sense amplifier is set to logic “0”; i.e., pre-charge phase. When the clock input in the sense amplifier changes to logic “1”; i.e., evaluate phase, the outputs of the sense amplifier are complementary. As shown in Figure 6.13, the PCSA detects the two complementary outputs of MTJs resistance after a delay of 45 ps during the evaluate phase. The N-MOS transistor M_{N0} acts as a sensing transistor, which senses the logic in the two complementary transistors and its width should be adjusted such that it senses logic values for both the MTJs.

6.1.1.3.1.2 Switching (Writing) Circuit A bi-directional current generator [38] is required in hybrid MTJ/CMOS logic designs to change the spin direction in the storage layer of MTJ device. The sense amplifier discussed above allows the reading circuit with low power and small area, and as a result, the MTJ writing circuit dominates main power and surface of the whole hybrid MTJ/CMOS logic circuit. Since STT writing approach is used, a very low bi-directional current pulse is needed, which is less than 200 microamperes for long current pulses. Figure 6.14 represents the switching circuit used for generating bi-directional currents. It consists of two NMOSs and two PMOSs configured in a way that two of these four transistors are always closed.

6.1.1.3.1.3 Logic Circuit Design Using MTJ/CMOS Hybrid Architecture There are two ways to design logic circuits using MTJs: one is discussed above, using hybrid CMOS/MTJ designs, whereas the other method is using the only MTJ device. As shown in Figure 6.11, for hybrid MTJ/CMOS logic, a PCSA is needed

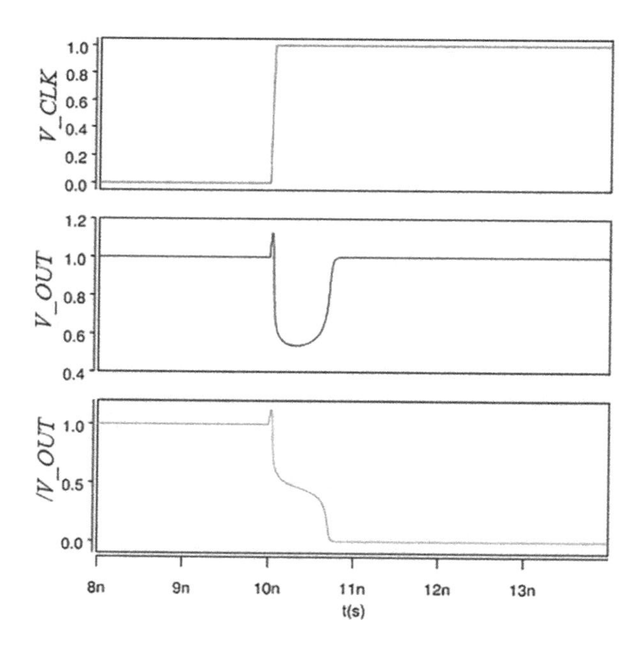

FIGURE 6.13 Simulation of PCSA using two complementary MTJs to detect the MTJ resistance states.

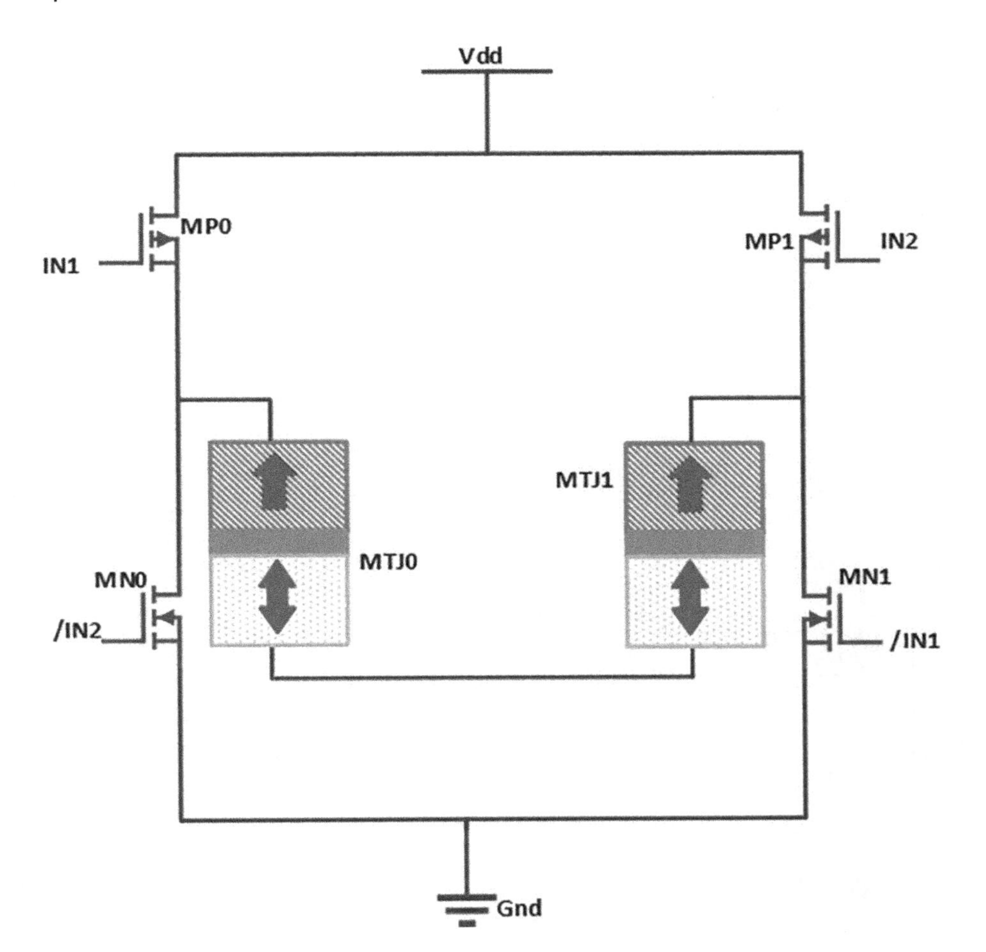

FIGURE 6.14 Switching (writing) circuit of MTJ for producing bi-directional current [38].

along with complementary MTJ and NMOS logic. The basic two-input NAND/AND logic using hybrid MTJ/CMOS logic as shown in Figure 6.15. For the circuit shown in Figure 6.15a, two outputs Out and Out_bar are obtained which give NAND and AND logic functions, respectively. The AND logic is formed such that the NMOS with input A and MTJ with input B are connected in series such that output $Out = AB$, whereas the other output is NAND from the series connections of NMOS and MTJ which has only $A\bar{B} + \bar{A}B$ logic with $\bar{B}A$ and BA missing (all these three connections give the output logic "1" for NAND gate). These missing connections to the above circuit cannot be formed as the architecture does not allow it. But still the correct NAND logic function is obtained by properly adjusting the widths of sensing NMOS transistors connecting the bottom electrodes of MTJs to sense the resistance of the two MTJ sub-branches. It is noteworthy that these missing connections have no impact on the logic. Table 6.4 presents the truth table for resistance configuration of NAND/AND logic. R_L and R_R are the two resistances

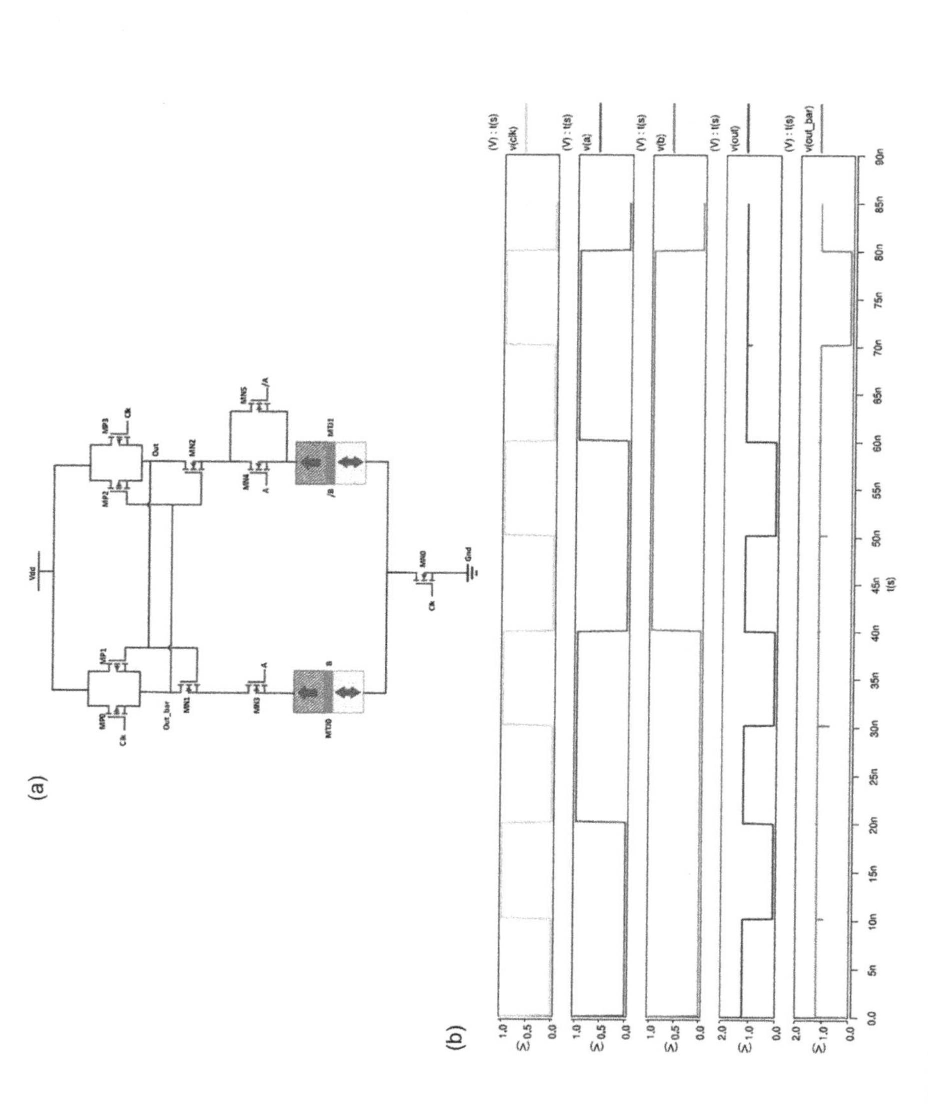

FIGURE 6.15 (a) Hybrid MTJ/CMOS NAND/AND logic gate. (b) Simulation results of the hybrid MTJ/CMOS NAND/AND logic gate.

TABLE 6.4
Resistance Configuration of NAND/AND Hybrid MTJ/CMOS Circuit

	B	Resistive Switching	OUT	Sub-Branch R_L	Sub-Branch R_R
0	0	$R_L < R_R$	1	R_{OFF}	$R_{OFF} + R_{ON}$
0	1	$R_L < R_R$	1	R_{OFF}	$R_{OFF} + R_{ON}$
1	0	$R_L < R_R$	1	R_{ON}	$R_{OFF} + R_{ON}$
1	1	$R_L > R_R$	0	R_{ON}	$R_{OFF} + R_{ON}$

of MOS transistors of two sub-branches of NMOS transistor with input A at the left and NMOS transistors connected in parallel with inputs "A" and "/A" or "$\bar{A}$" at the right sub-branch of the circuit. Assuming that $R_{OFF} > R_{AP}$, the PCSA senses the output accordingly. Figure 6.15b presents the simulation results of the hybrid MTJ/CMOS NAND/AND logic gate. By changing the parameters of the above circuit, we can convert the NAND/AND gate to NOR/OR or XOR/XNOR gates as shown in Figures 6.16a and 6.17a, respectively, while their simulation results are shown in Figures 6.16b and 6.17b, respectively.

6.1.1.3.1.4 Implementation of 1-Bit MTJ/CMOS Hybrid Adder (MFA) Figure 6.18 shows one-bit magnetic full adder where the sum logic has been designed as per Equation (6.3) and the carry logic has been designed as per Equation (6.4). This figure shows that the output carry logic consists of sub-branches "AC" and "$\overline{AC}$" which cannot be connected directly since they have no impact on the output by comparing inputs A and C. If these two inputs are different, then these two sub-branches will have the same resistance, but if they are same, then there will be two different resistances in those sub-branches, namely, R_L and R_R in the condition that $R_{OFF} > R_{AP}$. The simulation results of the magnetic full adder are shown in Figure 6.19.

$$\text{Sum} = \text{A } \mathbf{XOR} \text{ B } \mathbf{XOR} \text{ C} = \text{ABC} + \bar{\text{A}}\bar{\text{B}}\text{C} + \text{A}\bar{\text{B}}\bar{\text{C}} + \bar{\text{A}}\text{B}\bar{\text{C}} \tag{6.3}$$

$$C_o = \text{AB} + \text{BC} + \text{CA} \tag{6.4}$$

The comparison results in terms of power and delay of the designs are depicted in Figure 6.20. The comparison results suggest that the power efficiency and speed of MTJ-based design architecture is better comparing with conventional CMOS architecture. Considering simple NAND circuit, there is almost threefold decrease in power consumption and twofold increase in speed while switching from conventional CMOS architecture to only-MTJ architecture.

6.2.2 Spin Field-Effect Transistor

Spintronics has emerged as one of the most researched and emerging fields over the past two to four decades. The main aim of the field of spintronics is that the spin polarization is harnessed for the information processing and storage devices.

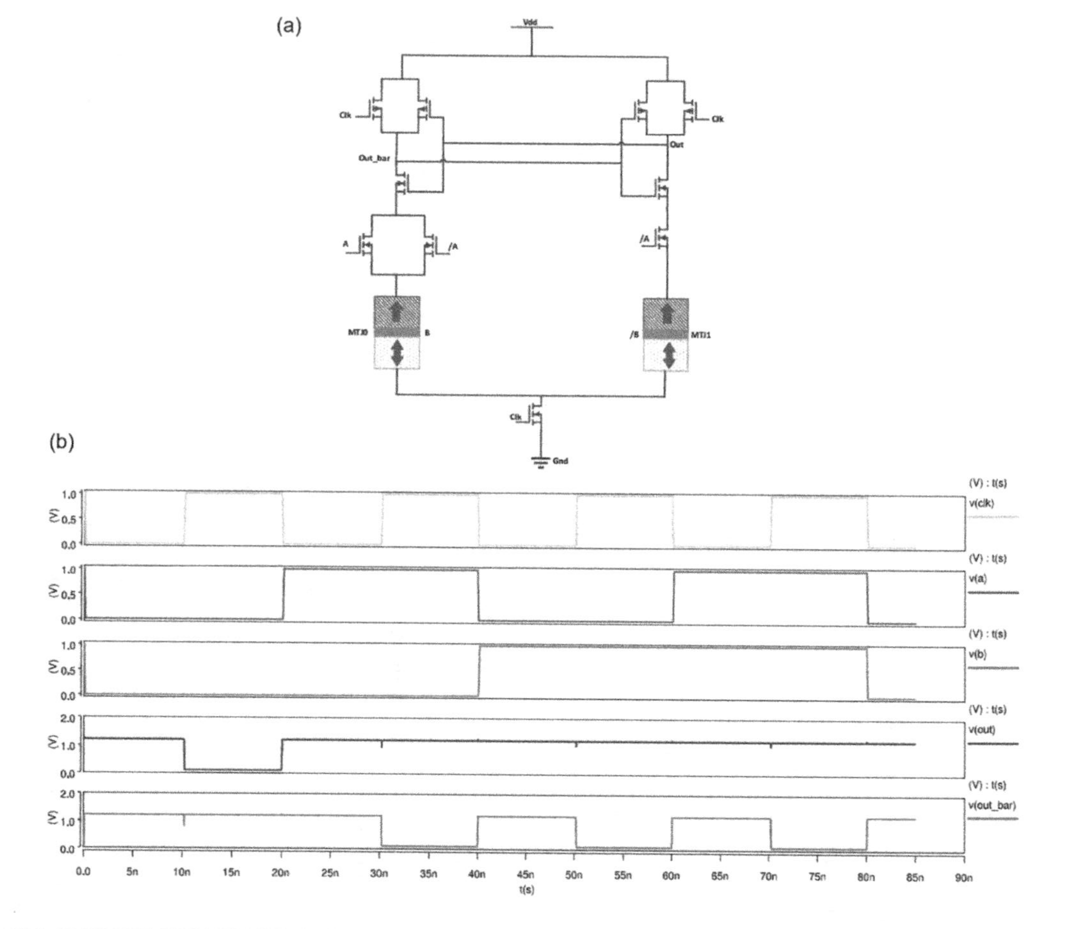

FIGURE 6.16 (a) Hybrid MTJ/CMOS NOR/OR logic gate. (b) Simulation results of the hybrid MTJ/CMOS NOR/OR logic gate.

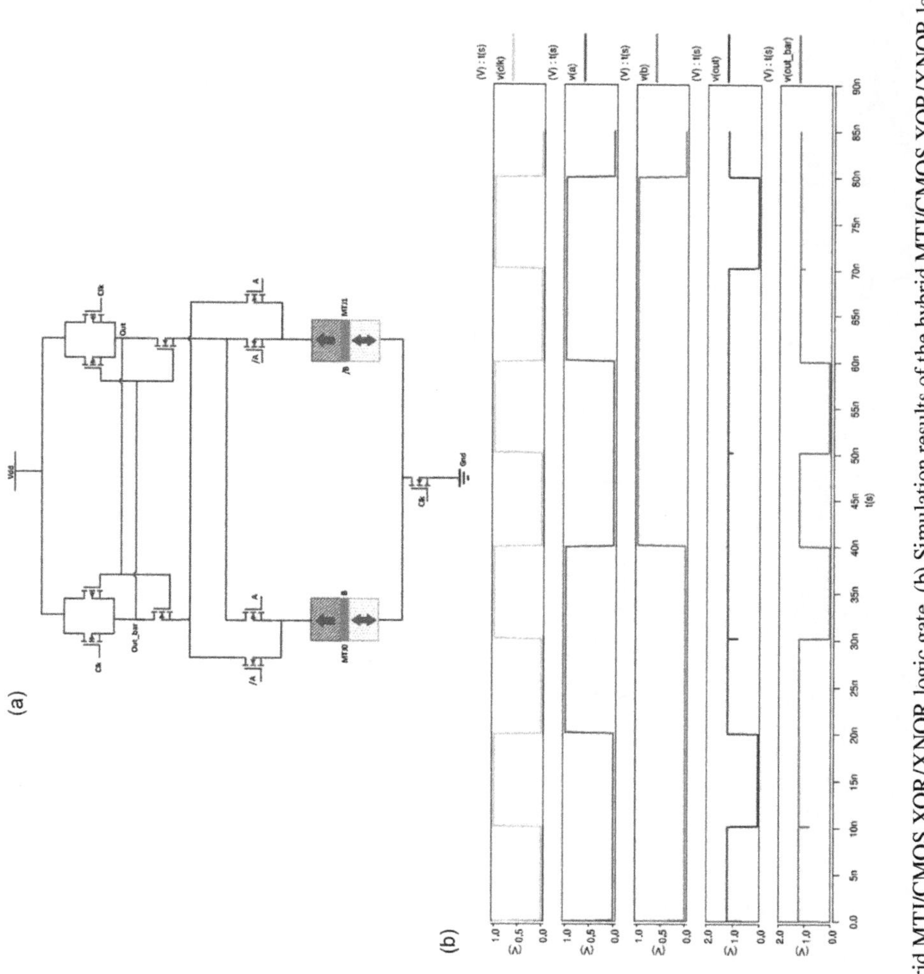

FIGURE 6.17 (a) Hybrid MTJ/CMOS XOR/XNOR logic gate. (b) Simulation results of the hybrid MTJ/CMOS XOR/XNOR logic gate.

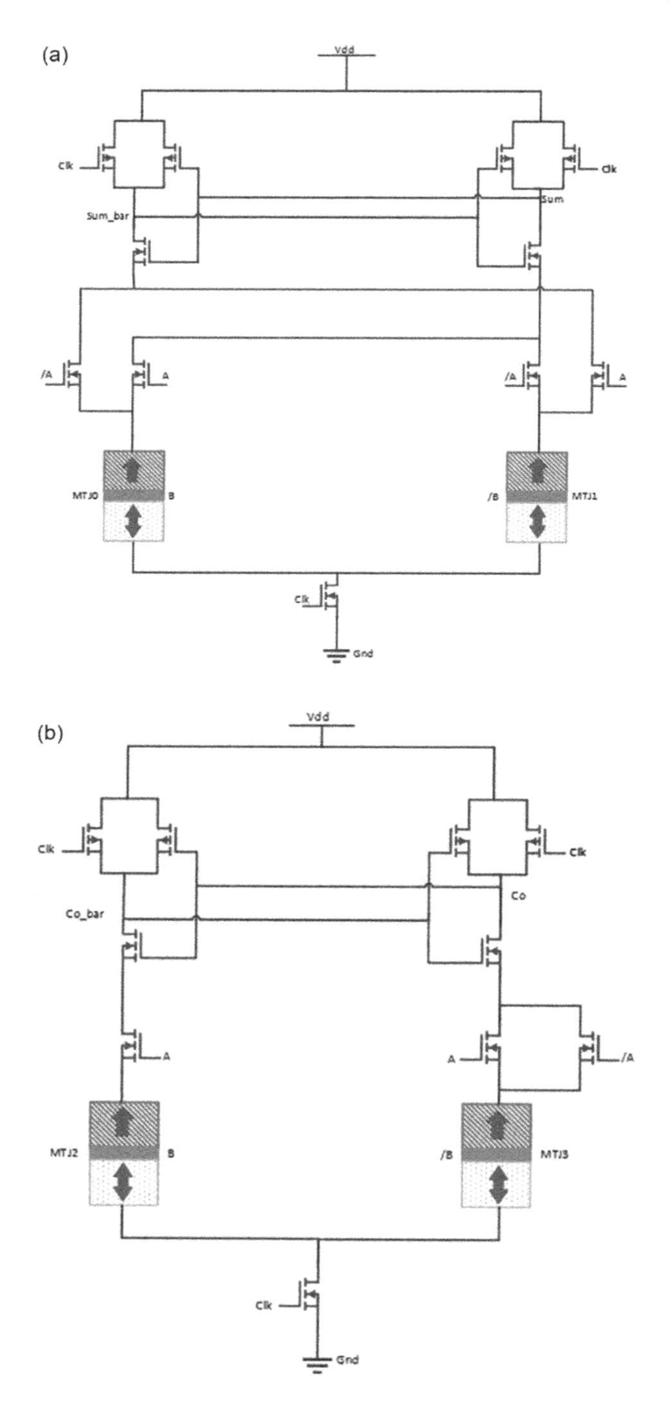

FIGURE 6.18 One-bit hybrid CMOS/MTJ Full adder [38], (a) Sum circuit, and (b) carry circuit.

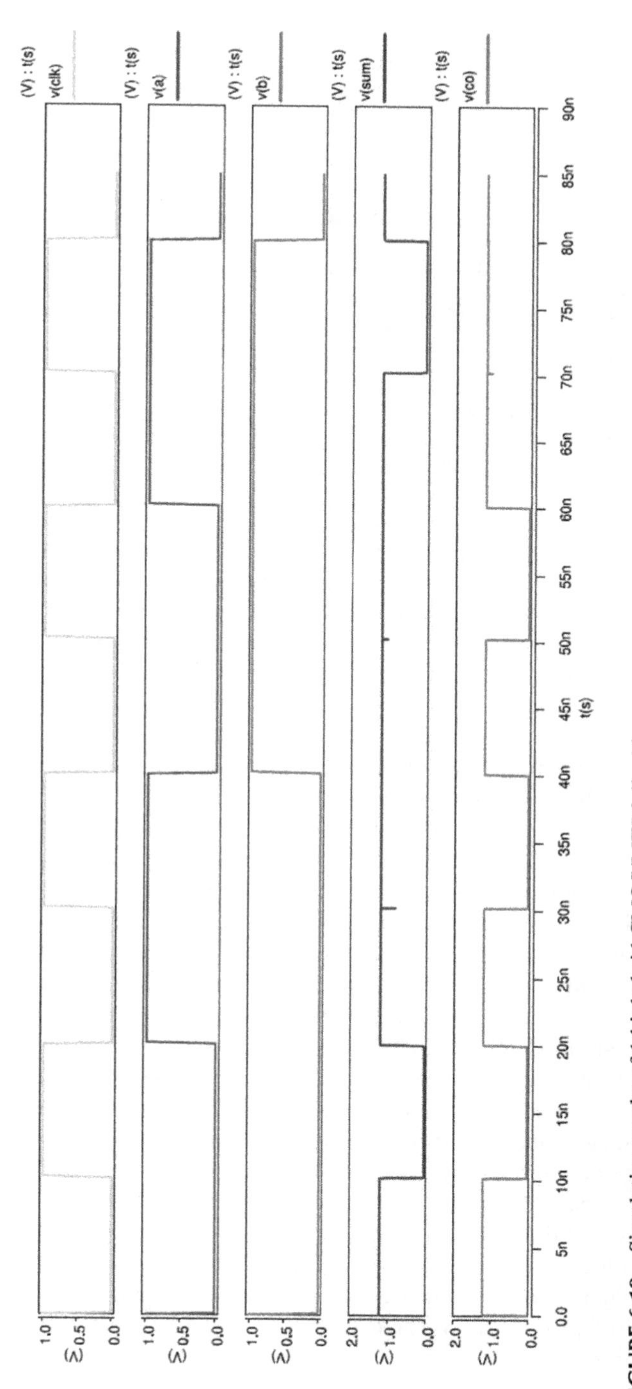

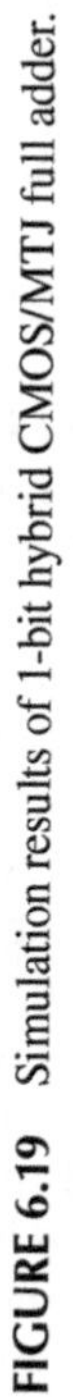

FIGURE 6.19 Simulation results of 1-bit hybrid CMOS/MTJ full adder.

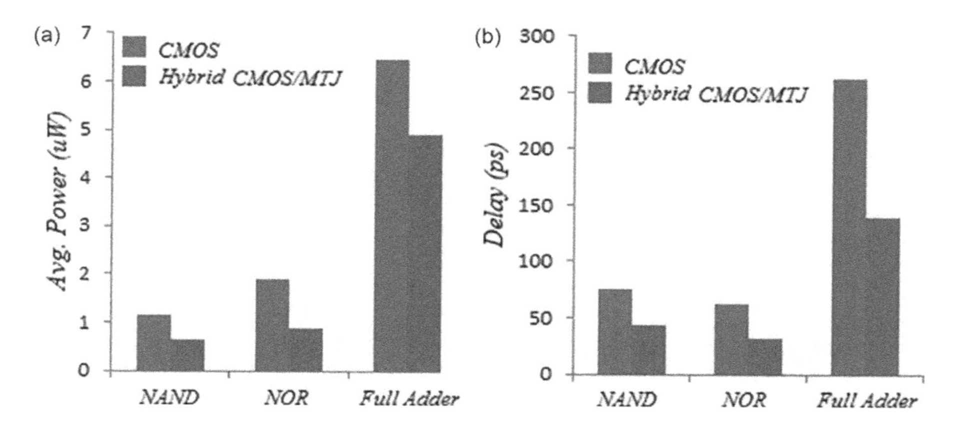

FIGURE 6.20 (a) Comparison of average power consumption for hybrid MTJ/CMOS NAND, NOR, and full adder circuits with their CMOS counterparts. (b) Comparison of delay consumption for hybrid MTJ/CMOS NAND, NOR and full adder circuits with their CMOS counterparts.

In conventional electronics, the charge of electrons is used for the process and flow of information, which has various disadvantages, i.e., mainly power consumption and speed degradation. It has been observed that while processing the spin polarization of electrons, the power required to flip the spin of an electron is very less as compared to the actual movement of electron from one place to another. This increases the capability of reducing power consumption as well as the speed. Based on this fact, as discussed earlier as well, various spintronic devices have been reported in the open literature wherein the spin polarization is exploited in place of the charge. The devices which are being reported in the literature are not fully spintronic devices because in addition to the flipping of the spin for changing its state, the actual movement of electrons is also considered and hence they are named as hybrid spintronic devices or simply the spintronic devices. The most important spintronic devices include MTJ (which was discussed in earlier sections) and Datta–Das transistor.

The Datta-Das transistor (usually called as spin field-effect transistor or simply spin-FET), which in general and specific forms is shown in Figure 6.21, is analogous to the conventional MOSFET with the only change being its drain and source electrodes are made of FM materials such as iron, cobalt, and nickel, while in case of the conventional MOSFETs, it is made of semiconducting material. Here, a non-magnetic material usually a semiconductor is being sandwiched between the FM electrodes. Advantages of using FM electrodes in the spin-FET include the elimination of the drain and source capacitances, which in turn reduces the power consumption, and hence addresses the power issues of devices. The other advantage of using FM source and drain is that they are used as spin polarizer and filter, respectively. The FM source injects the spin of single polarization, and the electrons pass through the non-magnetic channel, while reaching into the drain terminal, the FM drain acts as a spin filter, which allows the electrons of only one spin polarization, rejecting all other polarizations. It is a normally ON device. As soon as the voltage on gate terminal is applied, it induces a magnetic field named as the Rashba field which changes

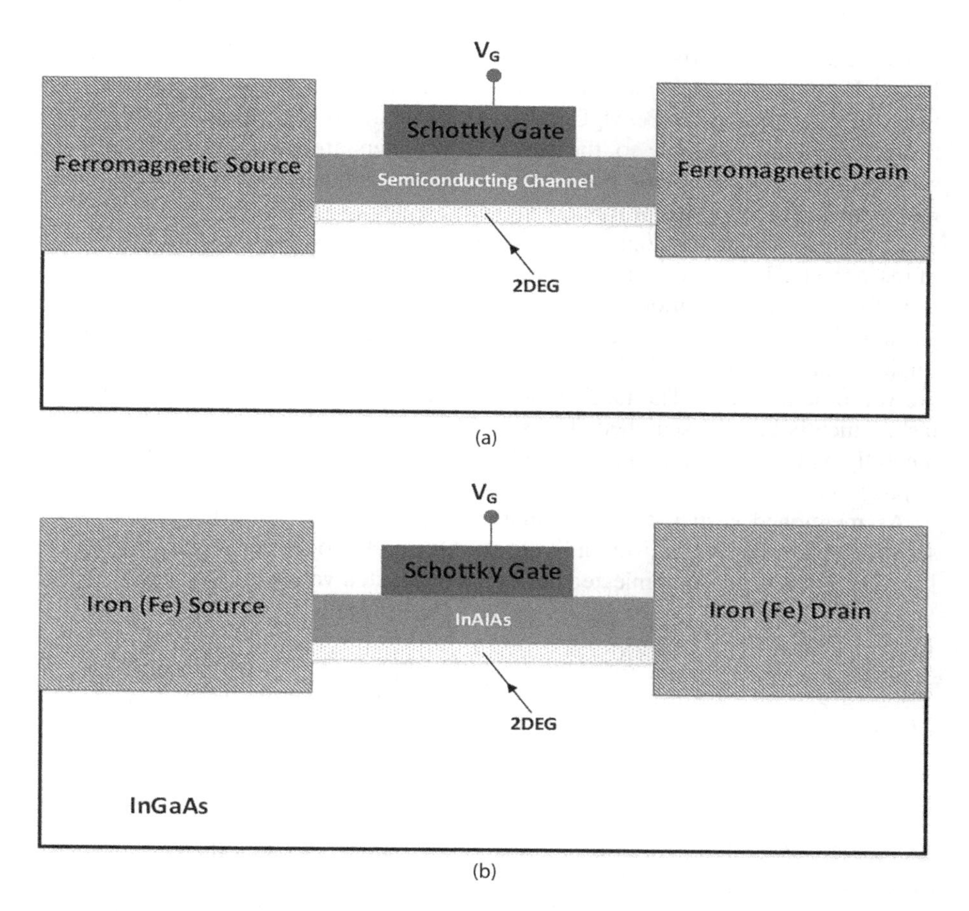

FIGURE 6.21 (a) General form of spin-FET. (b) Specific form of spin-FET [40].

the spin polarization of electrons in the channel, and hence, they are not accepted by the drain terminal. This way the spin-FET turns off.

The current research in the field of spintronic devices is eyeing the performance improvement of spin-FETs, wherein the researchers try to incorporate different materials and geometries to make these already existing devices more efficient to design improved circuits with them. Various techniques include the replacement of FM material by half-metallic materials such as Heusler alloys, CrO_2, Fe_2O_3, CrAs, MnAs, and CrSb [41–46], which increases the spin polarization efficiency of the devices. The spin-FET in which the electrodes are made of half metals is named as spin-MOSFET. The other thrust techniques include the usage of different emerging materials for channel such as graphene, metal chalcogenide, and 2D materials [47,48]. The change of geometries is also included wherein the gate engineering techniques are applied like the usage of multi-gates and gate all-around geometry [49]. SiO_2 is also being replaced with the emerging oxide materials like HfO_2. The MTJs are being improved by using various existing semiconductor and emerging materials as its scattering region like Fe/InAs/Fe or Co/graphene/Co and the interface between FM and semiconductor is made using oxides like MgO so the actual structure is a five-layered structure like Fe/MgO/InAs/MgO/Fe and Co/MgO/Graphene/MgO/Co, respectively [20]. By doing so, the various parameters are improved which include spin-injection efficiency, spin-polarization efficiency, and magnetoresistance (giant-magnetoresistance, GMR, or tunnel magnetoresistance, TMR).

The implementation of logic design (basic or higher) has also attracted the interest of researchers. The various simulated or mathematically validated devices are being used for the implementation of circuit design. Various circuits such as various logic gates and adders have been reported in the open literature, and their higher order implementations such as ALU (arithmetic and logic unit) and analog applications are yet to be designed. The fabrication of these spintronic devices is a challenging task, which is being researched. The spin-FET has just been demonstrated experimentally with limited spin injection and detection [50]. Its commercialization is still in progress.

As mentioned earlier, the mathematical models representing the operation of spin-FET have also been given in literature. One such model has been given in Ref. [51]. The relation between injected current and detected voltage is given as [49,52]

$$\frac{V_{\text{detect}}}{I_{\text{inject}}} = \pm 2R_N e^{\frac{L}{\lambda_N}} \prod_{i=1}^{2} \left[\frac{P_J \frac{R_i}{R_N}}{1-P_J^2} + \frac{P_F \frac{R_F}{R_N}}{1-P_F^2} \right] \times \left[\prod_{i=1}^{2} \left(1 + \frac{2\frac{R_i}{R_N}}{1-P_J^2} + \frac{2\frac{R_F}{R_N}}{1-P_F^2} \right) - e^{-\frac{2L}{\lambda_N}} \right]^{-1} \tag{6.5}$$

where "+" or "−" sign indicates the direction of magnetization of the two FM electrodes; R_F and R_N are the FM electrode resistance and semiconductor channel resistance, respectively; R_1 and R_2 are the interface resistances of the two junctions; P_J gives the polarization current of interfaces which reflects the capacity of polarization of junctions; P_F is the polarization current of FM electrode which reflects the capacity of FM electrode; L is the semiconductor channel length; and λ_N is the spin diffusion length in semiconductor channel (non-FM material), which is determined by the property of channel material.

For both contacts to be tunneling junctions such that it can generate the maximum spin signal, it is assumed that values of R_1 and R_2 are much larger than those of R_N. The expression therefore can be further simplified as

$$\frac{V_{\text{detect}}}{I_{\text{inject}}} = \pm \frac{1}{2} P_j^2 R_N e^{-\frac{L}{\lambda_N}} \tag{6.6}$$

Due to the applied voltage on gate terminal, the spin precession of injected electrons takes place and the spin precession angle is given as

$$\Delta\Theta = \frac{2m^* \alpha L}{\hbar^\circ} \tag{6.7}$$

where α gives the coefficient of spin orbit coupling, m^* gives the effective mass of an electron, and $\hbar$ is the reduced Planck's constant.

Spin orbit coupling factor depends upon the gate voltage as

$$\alpha = k_0 V_G^2 + k_1 V_G + k_2 \tag{6.8}$$

where k_0, k_1, and k_2 are fitting correlation constants.

After combining this model with the theory of Datta–Das [40], the following equations can be written to describe the behavior of spin-FET.

$$V_{\text{out}} = S_{\text{type}} \cdot I_{\text{inject}} n_s \cdot \cos\left(\frac{2m^* \alpha L}{\hbar^{o2}} + \varphi\right) \tag{6.9}$$

$$n_s = \frac{V_{\text{detect}}}{I_{\text{inject}}} = \pm\frac{1}{2} P_j^2 R_N e^{-\frac{L}{\lambda_N}} \tag{6.10}$$

$$\alpha = k_0 V_G^2 + k_1 V_G + k_2 \tag{6.11}$$

6.2.2.1 Multi-Gate Spin Field-Effect Transistor

According to International Technology Roadmap for Semiconductors [53], for next-generation industry, the alternative technologies and device structures need to be researched, so that high integration density, ultrafast switching and low power consumption [54], and one such promising technology is being researched using a spintronic device, spin-FET [55–57]. A novel device called multi-gate spin-FET is proposed where multiple gate voltages modulate the spin of electrons moving through the channel [58]. The general form of a multigate spin-FET is shown in Figure 6.22. The mathematical model of the multigate spin-FET can be obtained in the same way as for single-gate spin-FET. However, the impact of multiple gate voltages needs to be incorporated as

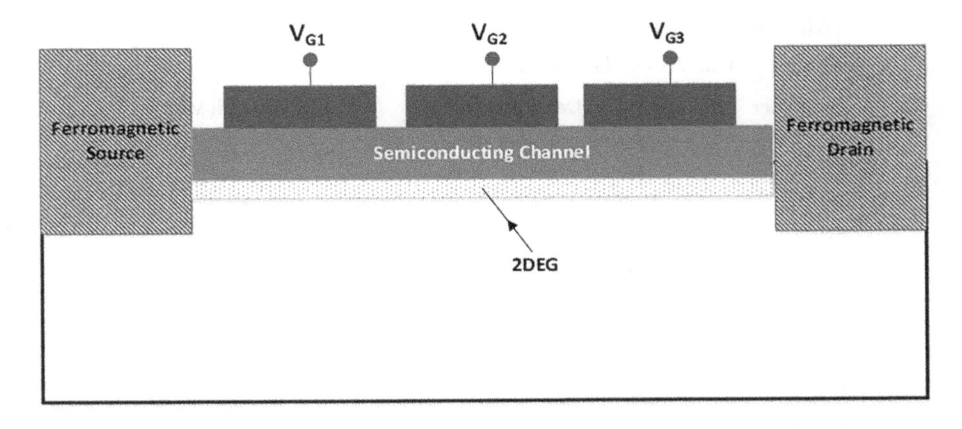

FIGURE 6.22 General structure of trigate spin-FET [40,49,52].

$$\Delta\Theta = \Delta\Theta_1 + \Delta\Theta_2 + \Delta\Theta_3 = \frac{2m^*(\alpha_1 + \alpha_2 + \alpha_3)L}{\hbar^\circ} \quad (6.12)$$

$$\alpha_i = k_0 V_{Gi}^2 + k_1 V_{Gi} + k_2 \quad (6.13)$$

Therefore, expression for output voltage can be written as

$$V_{out} = S_{type} \cdot I_{inject} n_s \cdot \cos(\Delta\Theta_1 + \Delta\Theta_2 + \Delta\Theta_3 + \varphi) \quad (6.14)$$

For the model parameters given in Table 6.5, the DC simulation results corresponding to Equation (6.13) for trigate spin-FET with $L = 600$ nm and $I_{inject} = 3\ \mu A$ are given in Figure 6.23.

6.2.2.2 Spin-FET-Based Logic Design

As can be noticed from Figure 6.23, the output peaks and valleys are changing with the given input voltage values indicating the truth table formation of the obtained logic function. The dc simulation for $V_{G1} = 1$, $V_{G2} = 0$, and $V_{G3} = 0$ and channel length varied from 600 to 1,200 nm is shown in Figure 6.24. Figure 6.24 shows that changing the channel length has significant effect on the output characteristics of the device; that is, the value of output voltage changes as the channel length is varied although the voltages at the three gates are kept the same. This feature can be exploited to design different logic gates from a single multigate spin-FET as shown in Figure 6.25, where logic functions of XOR, XNOR, and majority gate have been obtained for channel lengths of $L = 800$, 600, and 1500 nm, respectively [52].

6.2.2.3 Spin-FET-Based Reconfigurable Logic Design

The feature of reconfigurability is inherent in the spin-FET. This can be understood from the DC analysis of spin-FETs. The DC simulation results of trigate spin-FET are shown in Figure 6.26. The simulation results have been obtained by keeping

TABLE 6.5
Model Parameters Used in TGSF

Parameter	Description	Default Value
m^*	Effective mass of electron	$0.05 \times 9.1 \times 10^{-31}$ kg
$\hbar$	Reduced Planck's constant	1.054×10^{-34} J s
P_J	junction polarization	50%
R_N	Channel resistance	0.5 kΩ
λ_N	Spin diffusion length	1 μm
S_{type}	Type of spin-FET	1
φ	Phase shift correction	0
k_0	Fitting coefficient	0
k_1	Fitting coefficient	-3.4205×10^{-31}
k_2	Fitting coefficient	9.5775×10^{-31}

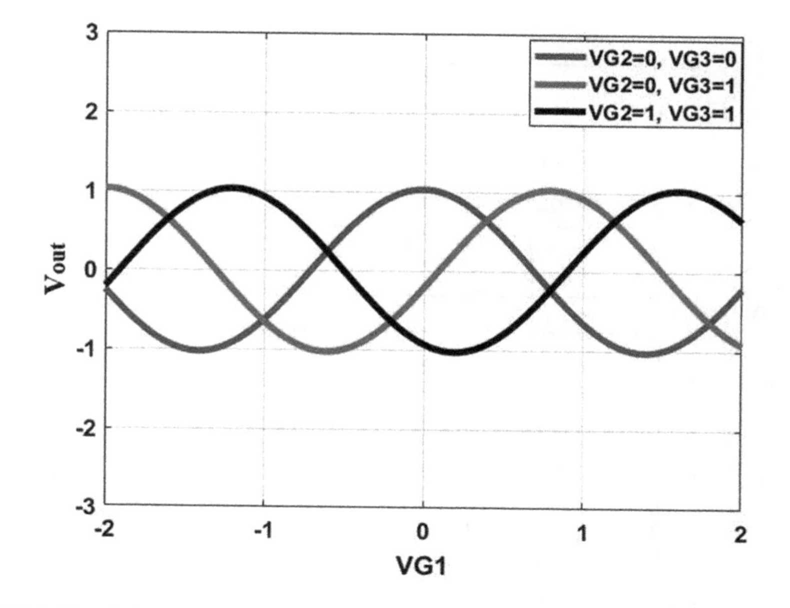

FIGURE 6.23 DC analysis of trigate Spin-FET with different voltages at three gates.

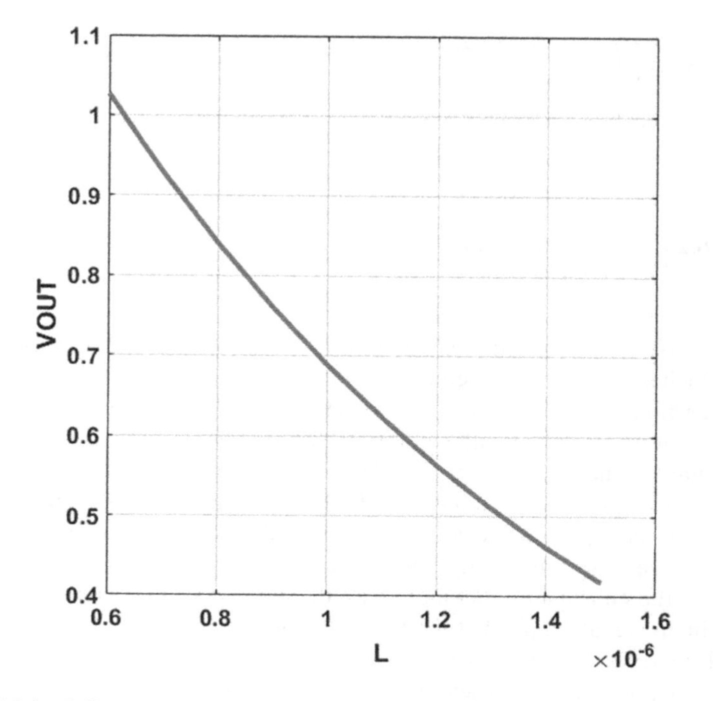

FIGURE 6.24 DC analysis of TGSF with different channel lengths.

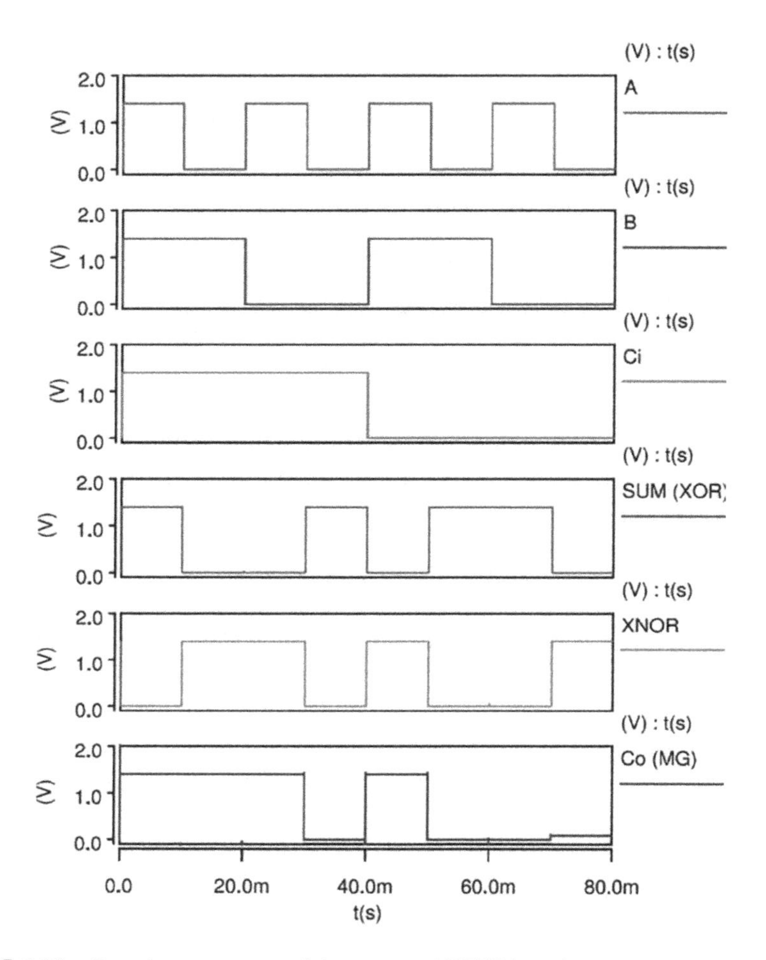

FIGURE 6.25 Transient responses of the proposed TGSF-based gates.

voltage at first terminal fixed at 1 V and using second gate terminal as control terminal on which the voltage is varied from 0 to 5 V with an increase of 1 V. The voltage at third terminal is varied analogously from 0 to 5 V. From Figure 6.26, it can be observed that the peak and valley points of different output signals change with different voltage at the control terminal. This feature can be exploited to obtain different logic functions from the same gate [49]. Further, the results show repeated peaks and valley as the input voltage at the gate terminal is varied; therefore, it is expected that the device shall repeat the same functions at different control voltages as all inputs have the same nature. In Figure 6.27, the first and second waveforms represent the two-input signals applied at the two different gate terminals, whereas the third to eighth waveforms represent 6 two-input symmetric functions corresponding to V_{G3} = 0–5 V (the *Y*-axis represent the voltage levels of input and output signals, and

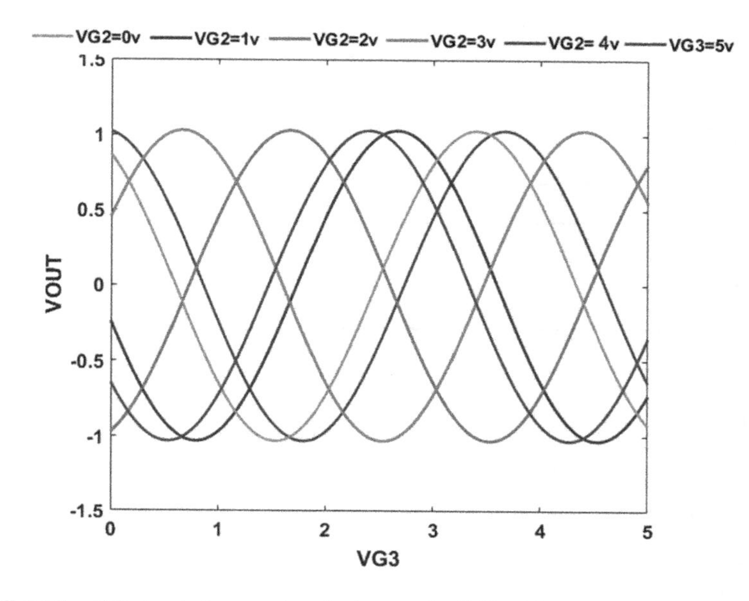

FIGURE 6.26 DC simulation results of trigate spin-FET with V_{G1} at 1V.

the *X*-axis represents the time interval). It can be noted that the XOR and XNOR functions are repeated at control voltages of 3 and 5 V. However, still four different two-input logic functions are obtained out of the six cases, which is a significant amount of reconfigurability.

6.3 SUMMARY

The exponential growth of semiconductor logic and memory devices over the past few decades has been possible by the fast scaling of CMOS technology. On the other hand, the developmental CMOS scaling has resulted in physical constraints, and the progress of CMOS technology therefore is expected to stop in the near future. As the physical gate length of CMOS device is getting closer to the physical constraints, numerous short channel effects emerge, bringing about high device leakage and performance instability, which enormously decay the energy efficiency and usefulness of CMOS circuits. Thus, the scaling down of device size will not be the solution for increasing computational power, but innovative switches (transistors) are to be found for further development in sub-10 nm range. Spin-based computing or spintronics is believed to be one of the promising technologies of the future. It besides the features offered by the CMOS technology offers the feature of non-volatility which can be well exploited for logic-in memory or bio-inspired computing. In this chapter, some discussions about the spintronics have been introduced. Besides, the different types of spin-based devices were discussed. In addition, the novel logic applications of these spin-devices were also presented.

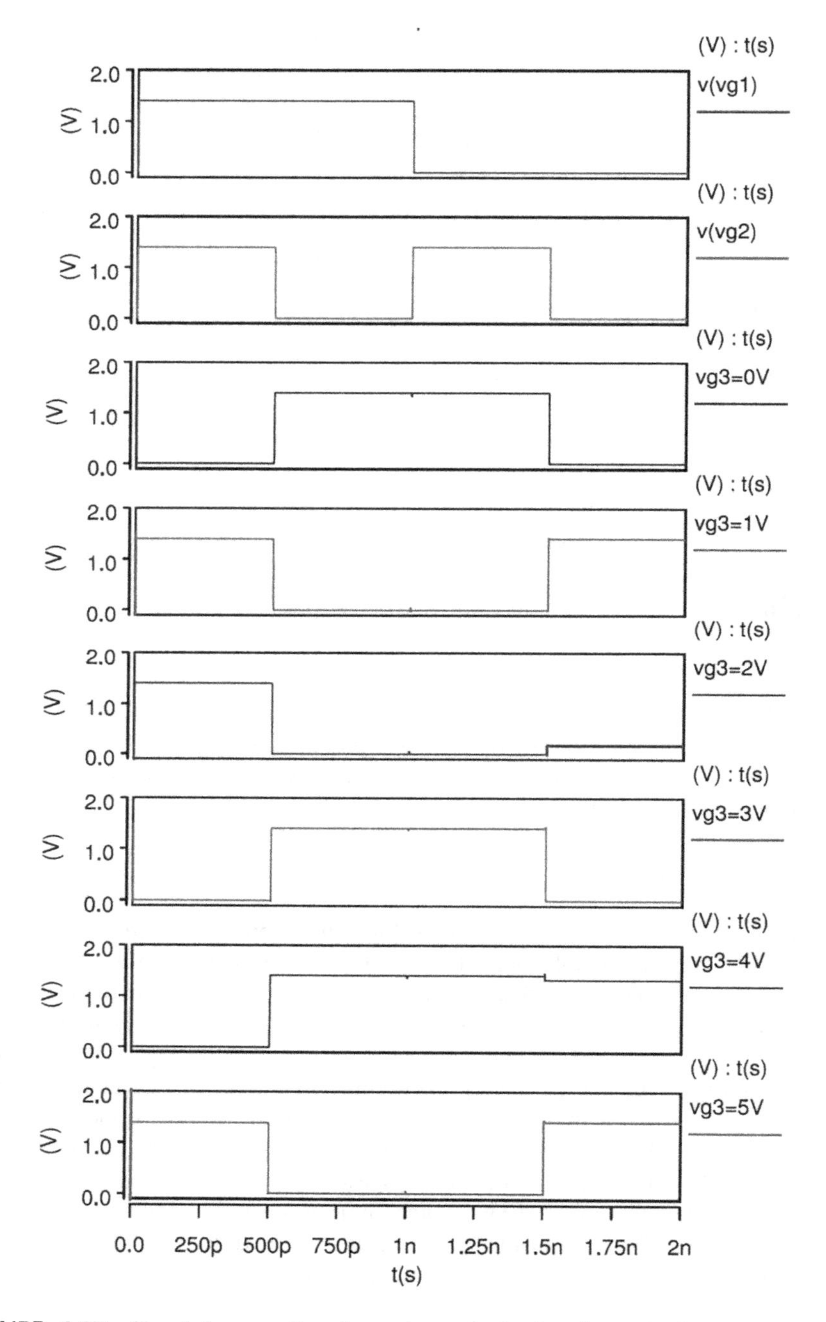

FIGURE 6.27 Simulation results of two-input logic functions obtained from trigate spin-FET.

REFERENCES

1. S. Bandyopadhyay, M. Cahay, *Introduction to Spintronics*, CRC Press Taylor & Francis Group (New York, NY, 2015).
2. E.R. Weber, *Semiconductors and Semimetals*, Academic Press, Elsevier (Burlington, MA, 1993).
3. P.P. Freitas, F. Silva, N.J. Oliveira, L.V. Melo, L. Costa, N. Almeida, *Sens. Actuators A*, 81, 2 (2000).
4. J. Wang, H. Meng, J.P. Wang, *J. Appl. Phys.* 97, 100509 (2005).
5. P.P. Freitas, L. Costa, N. Almeida, L.V. Melo, F. Silva, J. Bernardo, C. Santos, *J. Appl. Phys.* 85, 5459 (1999).
6. W.J. Ku, P.P. Freitas, P. Compadrinho, J. Barata, *IEEE Trans. Magn.* 36, 2782 (2000).
7. M.N. Baibich, J. M. Broto, A. Fert, F. Nguyen Van Dau, F. Petroff, P. Eitenne, G. Creuzet, A. Friederich, J. Chazelas, *Phys. Rev. Lett.* 61, 2472 (1988).
8. J.S. Moodera, L.R. Kinder, T.M. Wong, R. Meservey, *Phys. Rev. Lett.* 74, 3273–3276 (1995).
9. M. Julliere, *Phys. Lett. A* 54(3), 225–226 (1975).
10. B. Behin-Aein et al., *Nat. Nanotechnol.* 5(4), 266–270 (2010).
11. M. Sharad et al., “Ultra low energy analog image processing using spin based neurons”, *2012 IEEE/ACM International Symposium on Nanoscale Architectures (NANOARCH)*, Amsterdam, Netherlands, 2012.
12. E.Y. Tsymbal, D. Pettifor, *Solid State Phy.* 56, 113–237 (2001).
13. E.Y. Tsymbal, D. Pettifor, *Solid State Phys.* 56, 113–237 (2001).
14. N.F. Mott, *Proc. R. Soc. London Ser. A Math. Phys. Sci.* 156, 368–382 (1936).
15. L. Chang, M. Wang, L. Liu, S. Luo, P. Xiao, arXiv: 1412.7691 (2014).
16. C. Reig, M.D. Cubells-Beltrán, D. Ramírez Muñoz, *Sensors* 9(10), 7919–7942 (2009).
17. C. Tsang, R. E. Fontana, T. Lin, D.E. Heim, *IEEE Trans. Magn.* 30(6), 3801–3806 (1994).
18. E.I. Rashba, *Phys. Rev. B* 62, R16267 (2000).
19. R. Feiderling, M. Klein, G. Reuscher, W. Ossau, G. Scmidt, A. Waag, L. Molemkamp, *Nature* 402, 787 (1999).
20. G. Salis, R. Wang, X. Jiang, R.M. Shelby, S.S.P. Parkin, S.R. Bank, J.S. Harris, *Appl. Phys. Lett.* 87, 262503 (2005).
21. S. Matsunaga, J. Hayakawa, S. Ikeda, K. Miura, H. Hasegawa, T. Endoh, H. Ohno, T. Hanyu, *Appl. Phys. Express* 1(9), 091 301-1–091 301-3 (2008).
22. C.J. Lin, S.H. Kang, Y.J. Wang, K. Lee, X. Zhu, W.C. Chen, X. Li, W.N. Hsu, Y.C. Kao, M.T. Liu, W.C. Chen, L. YiChing, M. Nowak, N. Yu, T. Luan, “45nm low power CMOS logic compatible embedded STT MRAM utilizing a reverse-connection 1T/1MTJ cell”, *2009 IEEE International Electron Devices Meeting (IEDM)*, pp. 1–4, 2009.
23. S. Yuasa, D.D. Djayaprawira, *J. Phys. D Appl. Phys.* 40, R337 (2007).
24. S. Ikeda, H. Sato, M. Yamanouchi, H. Gan, K. Miura, K. Mizunuma, S. Kanai, S. Fukami, F. Matsukura, N. Kasai, H. Ohno, *SPIN* 02, 1240003 (2012).
25. B.N. Engel et al., *IEEE Trans. Magn.* 41(1), 132–136 (2005).
26. W.J. Gallagher, S.S.P. Parkin, *IBM J. Res. Dev.* 50, 5–23 (2006).
27. “Magnetic tunnel junctions for spintronics: principles and applications,” in *Wiley Encyclopaedia of Electrical and Electronics Engineering*, J. Webster Ed. John Wiley & Sons, Inc, (Hoboken, NJ, 2014).
28. J.Z. Sun, D.C. Ralph, *J. Magn. Magn. Mater.* 320, 1227–1237 (2008).
29. H. Zhao et al., *J. Phys. D Appl. Phys.* 45, 025001 (2011).
30. V.M. Edelstein, *Solid State Commun.* 73(3), 233–235 (1990).
31. J.E. Hirsch, *Phys. Rev. Lett.* 83(9), 065001 (1999).
32. Z. Wang et al., *Phys. Status Solidi Rapid Res. Lett.* 9(6), 375–378 (2015).

33. W. Zhao, Y. Cao, "New generation of predictive technology model for a sub-45nm design exploration", *2006 7th International Symposium on Quality Electronic Design, ISQED '06*, pp. 6–590, 2006.
34. J. Harms, F. Ebrahimi, X. Yao, J.-P. Wang, *IEEE Trans. Electron. Devices* 57(6), 1425–1430 (2010).
35. W. Kang, Y. Ran, Y. Zhang, W. Lv and W. Zhao, *IEEE Tran. Nanotechnol.* 16, 387–395 (2017).
36. Z. Diao, A. Panchula, Y. Ding, M. Pakala, S. Wang, Z. Li, D. Apalkov, H. Nagai, A. Driskill-Smith, L.-C. Wang, E. Chen, Y. Huai, *Appl. Phys. Lett.* 90(13), 132508 (2007).
37. S. Lee, S. Lee, H. Shin, D. Kim, *Jpn. J. Appl. Phys.* 44, 4S (2005).
38. S. Matsunaga, J. Hayakawa, S. Ikeda, K. Miura, H. Hasegawa, T. Endoh, H. Ohno, T. Hanyu, *Appl. Phys. Express* 1(9), 091301 (2008).
39. W. Kang, E. Deng, J.-O. Klein, Y. Zhang, Y. Zhang, C. Chappert, D. Ravelosona, W. Zhao, *IEEE Trans. Magn.* 50(6), 1–5 (2014).
40. S. Datta, B. Das, *Appl. Phys. Lett.* 56(7), 665467 (1990).
41. R.A. de Groot, F.M. Mueller, P.G. van Engen, K.H.J. Buschow, *Phys. Rev. Lett.* 50, 2024 (1983).
42. A. Yanase, K. Shiratori, *J. Phys. Soc. Jpn.* 53, 312–317 (1984).
43. K. Schwarz, *J. Phys. F Metal Phys.* 16, L211 (1986).
44. M. Shirai, *Phys. E Amsterdam* 10, 143–147 (2001).
45. M. Shirai, *J. Appl. Phys.* 93, 6844 (2003).
46. K. Sato, H. Katayama-Yoshida, *Semicond. Sci. Technol.* 17, 367 (2002).
47. J. Du et al., *Nanoscale* 9, 17585–17592 (2017).
48. C. Si, J. Zhou, Z. Sun, *ACS Appl. Mater. Interfaces* 7, 17510–17515 (2015).
49. G.F.A. Malik, M.A. Kharadi, F.A. Khanday, *Microelectron. J.* 90, 278–284 (2019).
50. J. Nitta, T. Bergsten, Y. Kunihashi, M. Kohda, *J. Appl. Phys.* 105, 122402 (2009).
51. G. Wang, Z. Wang, J. O. Klein, W. Zhao, *IEEE Trans. Magn.* 53(11), 1 (2017).
52. G.F.A. Malik, M.A. Kharadi, F.A. Khanday, *IET Circuits Devices Syst.*, to appear in IET Circuits Devices and Systems (IET), DOI: 10.1049/iet-cds.2019.0329.
53. 2015 International Technology Roadmap for Semiconductors (ITRS). online. Available: https://www.semiconductors.org/
54. J. Shen et al., *Phys. Rev. Appl.* 6(6), 064028 (2016).
55. Z. Wang et al., *IEEE Electron Device Lett.* 39(3), 343–346 (2018).
56. X. Lin et al., *Phys. Rev. Appl.* 8(3), 034006 (2017).
57. W.Y. Choi, H.J. Kim, J. Chang, S.H. Han, H.C. Koo, *J. Electron. Mater.* 46(7), 3894–3898 (2017).
58. G. Wang, Z. Wang, X. Lin, J. OliverKlein, W. Zhao, *IEEE Trans. Magn.* 54(11), 2831696 (2018). DOI: 10.1109/TMAG.2018.2831696.

7 Phase-Change Devices and Their Applications

7.1 INTRODUCTION

Random access memory (RAM) can be either volatile or nonvolatile. A volatile memory loses its previous stored data on removing the power supply as is the case for dynamic random-access memory (DRAM) and static random-access memory (SRAM). For nonvolatile memory, the contents that were stored previously will continue to be retained even after the removal of the supply. Flash memory is a typical example of nonvolatile memory. Memory technologies combine the advantages and disadvantages to achieve higher performance. For example, DRAMs used in a computer system have high capacity and density, but they are volatile meaning there is a need to refresh every few milliseconds. Due to this refreshing, the energy consumption of the device increases which is not desirable. SRAM, on the other hand, is fast but is also volatile just like the DRAM; in addition, SRAM cells are of larger size which hinders its implementation on a large scale. Flash memory, which essentially consists of a metal oxide semiconductor field-effect-transistor (MOSFET) in addition to a floating gate in each memory cell, is currently being used extensively particularly for the embedded applications owing to its low cost and high density. Depending upon how memory cells are organized, flash memory is classified as NOR Flash and NAND Flash [1]. In NOR Flash, cells are read and programmed individually as they are connected in parallel to bit lines. This resembles the parallel connection of transistors in a CMOS NOR gate architecture. For the case of NAND Flash, the architecture resembles that of a CMOS NAND gate as the cells are connected in series to the bit lines. It must be noted that less space is consumed by the series connection as compared to the parallel one which results in a reduced cost of NAND Flash. However, both types of flash memories have several disadvantages such as low operation speed (write/erase time: 1 ms/0.1 ms), limited endurance (10^6 write/erase cycles), and high write voltage (>10 V) [2].

The memory technologies mentioned above, i.e., DRAM and SRAM, as well as flash, are charge storage-based memories. DRAM stores the information in the form of charge in the capacitor, SRAM is based on the storage of charge at the nodes of the cross-coupled inverters, whereas the flash memory technology uses the floating gate of the transistor to store the charge. All these existing charge storage-based memory technologies are currently facing challenges to scale down to 10 nm node or beyond. This is attributed to the loss of stored charge at nanoscale, which results in the degradation of the performance, reliability, and noise margin. In addition, requirements of large refresh dynamic power for DRAM and leakage power for both SRAM and DRAM pose serious challenges for the design of future memory hierarchy.

Therefore, a new class of memories usually referred to as emerging memory technologies are currently undergoing development and are being actively researched primarily in the industry with the aim to revolutionize the existing memory hierarchy [3]. These emerging memory technologies aim to integrate the switching speed of SRAM, storage density comparable to that of DRAM, and the non-volatility of flash memory, and thus become very attractive alternatives for future memory hierarchy.

To classify a memory device as an ideal one, it should have the following characteristics: low operating voltage (<1 V), long cycling endurance (>10^{17} cycles), enhanced data retention time (>10 years), and low energy consumption (fJ/bit), as well as superior scalability (<10 nm) [4]. However, there is no single memory to date that satisfies these ideal characteristics. Various emerging memory technologies are actively being investigated to meet a part of these ideal memory characteristics. These memory technologies that depend upon the change of resistance rather than charge to store the information are as follows: (1) phase-change memory (PCM), (2) spin-transfer torque magnetoresistive random-access memory (STT-MRAM), (3) memristor, and (4) resistive random-access memory (RRAM). In PCM, the switching medium consists of a chalcogenide material (commonly Ge_2-Sb_2-Te_5 or GST) [5–7]. PCM relies on the difference in resistance between the crystalline phase (LRS) and amorphous phase (HRS) for efficient data storage capability. For SET operation, PCM is heated above its crystallization temperature on the application of voltage pulse, while for RESET operation, a larger electrical current is passed through the cell and then abruptly cut off so as to melt and then quench the material in order to achieve the amorphous state. In STT-MRAM, the storage capability is due to the magnetic tunneling junction (MJT) [8–10], which consists of two ferromagnetic layers and a tunneling dielectric sandwiched between them. The magnetic direction of the reference layer is fixed, whereas the application of external electromagnetic field can change the magnetic direction of the free ferromagnetic layer. If the reference layer and the free layer have the same direction of magnetization, the MTJ is referred to be in the LRS. For MTJ, to be in the HRS, the direction of the magnetization of two ferromagnetic layers is antiparallel. Memristor involves the resistance change in material due to phase change [11]. Memristor has pinched hysteresis characteristics. Some of the applications of logic design using memristors have been presented in the literature. However, most of the applications of memristors fall in the domain of neuromorphic design [11]. RRAM consists of an insulating layer (I) sandwiched between the two metal (M) electrodes [12,13]. RRAM relies on the formation and the rupture of conductive filaments (CFs) corresponding to LRS and HRS, respectively, in the insulator between two electrodes [14–16]. RRAM can be also based on memristors, and such RRAMs are called memristor-based RRAMs. In fact, the difference between the memristor and RRAM is not quite clear in the literature, and the memristor and RRAM are used interchangeably. However, given their different domains of applications evident from the presentation of their characteristics, they have been considered in different sections in this chapter. A detailed comparison of existing and emerging memory technologies is shown in Table 7.1. As is evident from the table, STT-MRAM and PCM have smaller area as compared to SRAM. While STT-MRAM offers fast write/read speed, long endurance, and low

TABLE 7.1
Comparison of Emerging Memory Technologies

Memory Technology	SRAM	DRAM	NAND FLASH	NOR FLASH	PCM	STT-MRAM	RRAM/ Memristor
Cell area	>100 F^2	6 F^2	<4 F^2	10 F^2	4–20 F^2	6–20 F^2	<4 F^2(3D)
Cell element	6T	1T1C	1T	1T	1T(D)1R	1(2)T1R	1T(D)1R
Voltage	<1 V	<1 V	<10V	<10V	<3V	<2V	<3V
Read time	~1 ns	~10 ns	~10 ns	~50 ns	<10 ns	<10 ns	<10 ns
Write time	~1 ns	~10 ns	100 μs^{-1} ms	10 μs^{-1} ms	<50 ns	<5 ns	<10 ns
Write energy (J/bit)	~1 fJ	~10 fJ	~10 fJ	100 pJ	~10 pJ	~0.1 pJ	~0.1 pJ
Retention	N/A	~64 ms	>10years	>10years	>10years	>10years	>10years
Endurance	>10^{16}	>10^{16}	>10^4	>10^5	>10^9	>10^{15}	>10^6–10^{12}
Multibit capacity	No	No	Yes	Yes	Yes	Yes	Yes
Non-volatility	No	No	Yes	Yes	Yes	Yes	Yes
Scalability	Yes	Yes	Yes	Yes	Yes	Yes	Yes

F, feature size of lithography.

programming voltage. PCM has a disadvantage of extensive write latency. RRAM has lower programming voltage and faster write/read speed compared to flash and is seen as potential replacement of flash memory. Among all the emerging memory technology candidates, RRAM has significant advantages such as easy fabrication, simple MIM structure, excellent scalability, nanosecond speed, long data retention, and compatibility with the current CMOS technology, thus offering a competitive solution to future digital memory [17].

7.2 PHASE-CHANGE MEMORY (PCM)

The PCM is an emerging non-volatile semiconductor technology. The basic phenomenon in PCM involves the resistance change in thin-film chalcogenide material induced by thermal phase transition [18–21]. The chalcogenide material is the combination of the elements belonging to the oxygen group. The GST is found to be the efficient chalcogenide material, which exhibits the two stable states: crystalline and amorphous, having the large resistivity contrast [22]. The transition between the two phases can be achieved electrically. The crystalline phase has the small resistance and can be considered as the binary 1 (SET), whereas the amorphous phase has the large resistance and can be considered as the binary 0 (RESET). The physics of the phase-change material is based on Joule's law of heating. When a large amount of current is given, the heat gets produced, corresponding to the melting point of the material, and the material loses its structure and becomes amorphous in nature. This high current is given for small amount of time so that the material may not form any regular structure, whereas when the specific current is given for a large amount of time that produces the heat corresponding to the temperature between the transition and melting point of the material, the material forms a regular structure. In between crystalline and amorphous phases, there exist a large number of intermediate resistance levels that are used to store multiple levels in multi-level cell (MLC) [5]. PCM technology is considered as an alternative to mainstream flash technology due to its maturity for large volume manufacturing, good electrical performances, superior reliability, and bit granularity. In PCM field, the primary focus of the researchers has been to reduce the programming current and to optimize the device for different requirements and applications.

7.2.1 Overview of Phase-Change Material Properties and Chalcogenide Materials for PCM

Materials containing S, Se, or Te alloyed with usually group 15 (Sb, As) or group 14 (Sn, Ge) elements are known as chalcogenide materials. The chalcogenide materials can change their state from amorphous form to the crystalline form possessing different optical and electrical properties. In 1969, Ovshinsky reported this reversible transition in chalcogenide materials contacting Si, Ge, Te, and As. $Ge_2Sb_2Te_5$ is the most studied chalcogenide phase-change material [18,23,24]. While changing from crystalline state to amorphous state, the electrical resistivity changes by three to four orders due to the suppression of resonant bonding in amorphous state. This forms the basis of using such materials in the PCMs.

To change the state, we need to apply electrical pulses. In order to change the state from crystalline state to amorphous state, a RESET pulse is applied where a large amount of current is given for short period of time. The SET pulse (a current pulse having small intensity than RESET pulse applied for large period of time) changes the state from amorphous form to crystalline form.

The properties of the phase-change devices actually depend on the phase-change material properties. Thus, material science is involved to understand the properties like the crystallization phenomenon of phase-change devices. Only for the well-defined composition as $(GeTe)_n(Sb_2Te_3)_m$ (where n and m are integers), the stable compounds of phase-change materials exist. In these materials, periodic stacking of planes is observed. These planes are perpendicular to threefold axis and are completely occupied by Te atoms. However, on the pseudo-binary line for ternary alloys, the phase is either rhombohedral or cubic obtained by the crystallization of amorphous phase. When the alloy is expressed as $(GeTe)_{1-x}$ $(Sb_2Te_3)_x$, the rhombohedral phase is obtained for lower values of x and the cubic phase is obtained for the values of x from 0.14 to 0.67. The cubic phase is transformed into hexagonal phase irreversibly by heating and is thus a meta-stable phase. $Ge_2Sb_2Te_5$ is the popular phase-change material that is widely used in resistive memories. Due to having good cyclability and fast crystallization, it is used in most applications. The phase-change materials show paradoxical characteristics. Their varying reflectivity and electrical conductivity of amorphous and crystalline phases suggest their bonding and structural differences, whereas at crystallization, the fast crystallization suggests minor arrangements of their atoms [18].

The total pair correlation function is used to describe the medium- and short-range atomic order in the amorphous materials. This function can be achieved by the Fourier transform of structure factor that can be measured either by X-ray total scattering or by neutron total scattering. In some cases, it can also be measured by using high brilliance or by using synchrotron radiation. On the pseudo-binary line, the cubic Ge-Sb-Te phases and $Ge_2Sb_2Te_5$ phases show a huge amount of disorder. Thus, their average structure differs from their local structure observed from the Bragg peak analysis in the experiments of X-ray diffraction. The AIMD (ab initio molecular dynamics) along with density function allows simulating the crystalline or amorphous states and measuring their electronic properties [18].

With the space group of R3m, the amorphous phase of GeTe crystallizes into rhombohedral phase, which is a stable phase. Due to the reversible transformation of its structure and ferroelectric character, it has been widely studied. The rhombohedral phase of GeTe is taken as distorted rock-salt structure, which along the diagonal of cube has undergone a shear. In such a case, the angle of unit cell is equal to 88.17° and its edge is equal to 0.5985 nm at 295 K. Ge is located off-centered in 0.525, 0.525, and 0.525 if Te atom is at origin. The ferroelectric properties of rhombohedral GeTe are due to Ge displacement. The six Te atoms surround the Ge giving rise to a distorted octahedron. The Te-Ge distances are split into three long (0.316 nm) and three short (0.284 nm) distances due to Ge off-centering. This splitting is due to the result of the Peierls distortion [25]. The coordination of Ge is represented as (3 + 3). The angle between long and short bonds is equal to 171.8°.

At the local scale, on the pseudo-binary line, the cubic phase of Ge-Sb-Te alloys have huge amount of disorder. This disorder does not exist in rhombohedral GeTe. However, both cubic and rhombohedral phases have distorted octahedral structure due to the existence of long and short Te-Ge bonds and the absence of homo-polar bonds. These features of structure need to be compared with local structure of amorphous phase in order to understand properties of phase-change materials [18].

In amorphous Ge-Sb-Te alloys, $Ge_2Sb_2Te_5$, and binary amorphous phase of GeTe, the local atomic order is similar. No short Te-Te bonds exist in GeTe. The same trend is observed in amorphous $Ge_1Sb_2Te_4$, $Ge_2Sb_2Te_5$, or $Ge_8Sb_2Te_{11}$. Te-Sb short bonds are also present in addition to Ge-Te and Ge-Ge short bonds in ternary alloys. By integrating the calculated partial pair distribution functions, coordination numbers of a given element can be found out, but at the same time, their values are dependent on range of integration. Nevertheless, the coordination number in the crystalline state of Ge-Sb-Te and GeTe is systematically higher than that in amorphous state. Ge atoms have the bond angle around 90° and are found to be either in distorted octahedral environment or in tetrahedral environment. However, in crystalline state only distorted octahedral environment exists. The octahedral environments in amorphous state are more distorted than those in crystalline state. In amorphous state, Ge and to somehow Sb displacements are larger. As a result, by comparison with crystalline phase, the long bonds expand and the short Te-Ge (Te-Sb) bonds shrink in amorphous state [18].

A larger volume reduction is exhibited by Ge-Sb-Te alloys on crystallization which results in mechanical stress and formation of voids that in turn limits the reliability and cyclability of PCM devices. Thus, in order to reduce volume shrinking, the light doping elements such as C, N, or O are added to increase the crystallization temperature.

In amorphous state, both GeTe and $Ge_2Sb_2Te_5$ are semiconductors, and thus, in the amorphous state, resistivity smoothly decreases with increase in temperature. An abrupt drop in resistivity takes place when GeTe is crystallized into rhombohedral phase and $Ge_2Sb_2Te_5$ is crystallized into cubic phase. The resistivity of GeTe on heating after crystallization becomes almost temperature independent. GeTe is degenerate type of semiconductor and is p-type, metal-like conductor because of large hole concentration. However, in cubic phase, the resistivity of $Ge_2Sb_2Te_5$ is reduced with increase in temperature after crystallization. The phase-change materials also exhibit different optical properties besides having different electrical properties in crystalline and amorphous state. The optical constants for hexagonal phases and crystalline phases of $Ge_2Sb_2Te_5$ are very similar. An optical band-gap of about 0.7 eV for amorphous phase and 0.5 eV for hexagonal crystalline phase is reported in $Ge_2Sb_2Te_5$. The film microstructure and deposition parameters determine the optical properties of phase-change materials. The phase-change materials also exhibit resonant bonding in crystalline state where in average, every atom possesses three p valence electrons and has to be bounded by six neighboring atoms covalently. Pauling stated that resonance between nonbonding and bonding state can occur giving rise to "resonance bonding." The crystal that is bounded resonantly shows high electronic polarizability and strong electron delocalization. This causes high refractive index and high optical dielectric constant [18].

The phase-change materials must also have data retention property. The retention time at a given temperature is defined as elapsed time before the crystallization of the amorphous state. Hence, the improvement of the amorphous state stability against crystallization is required. At operating temperature, the required data retention is 10 years. The retention time obeys the Arrhenius law as a function of inverse of temperature. Thus, proposals have been given to increase the crystallization temperature. Among them, one proposal is to add dopants (C, O, B, N, SiO2, SiN, or SiC) in GeTe and Ge-Sb-Te alloys in relatively high concentration. Mostly, N and C are added as impurities. As compared to the undoped $Ge_2Sb_2Te_5$, the undoped GeTe has better retention properties. The similar effects are observed in N- and C-doped alloys. For crystallization, with doping, the activation energy increases [18].

The resistivity of amorphous phase increases with time due to material aging by the phenomenon called as "resistance drift." This can cause failure of memory and is one of the major challenges of the multilevel storage devices. This resistance drift is related to the "structural relaxation" of amorphous phase. In the PCM cell, the resistance drift causes the increment in the resistance of the RESET state. In the crystalline state, the resistance drift is negligible in the SET state [18].

In PCM cells, the decrease in the programming current is the major issue. By optimizing composition of material like by doping, programming current could be decreased. Changing the architecture of memory cell is another approach. To improve the Joule heating efficiency and to decrease the RESET current, the phase-change materials are confined in a small structure. However, in such structures the phase-change materials remain in contact with insulators besides metallic electrodes. However, the crystallization of phase-change materials is affected by the cladding materials only for film thickness under 10 nm. In GeTe and $Ge_2Sb_2Te_5$, inside the volume of film, the nucleation occurs in the absence of oxidation, and at the oxidized upper surface, heterogeneous nucleation occurs. The onset of crystallization of the upper surface can be controlled by controlling the degree of oxidation. The crystallization of phase-change materials such as $Ge_2Sb_2Te_5$ and GeTe alloys can be selected by interface engineering. At material surface, the heterogeneous nucleation is eliminated in the absence of oxidation and inside the phase-change material volume, the nucleation occurs. As a result, in non-oxidized film, the crystallization temperature is higher, and hence by avoiding surface oxidation, the stability of amorphous phase in GeTe and $Ge_2Sb_2Te_5$ can be increased. This opens the way towards the architecture of new PCM cell with better data retention, provided no oxidation of phase-change material occurs during the process of fabrication. A detailed discussion on the materials used for PCM can be found in Ref. [18].

7.2.2 Scaling of Phase-Change Memory Devices

The scaling of PCM is based on its material properties. Several compositions of phase-change materials are studied to observe which material suits well for better performance. Scaling in one dimension means to reduce the thickness of phase-change material while keeping other parameters constant. As per the recent studies, with the decrease in thickness, the crystallization temperature increases. The variation of crystallization temperature with thickness satisfies the following empirical equation:

$$T_X = T_{ax} + (T_m - T_{ax})e^{\frac{-d}{C}} \tag{7.1}$$

where T_{ax} is the crystallization temperature of bulk material or thick film, T_m is the melting temperature, d is the thickness, and C is fitting constant. The Gibbs free energy of system can be related by the following empirical equation:

$$T_m = T_{ax}\left(\frac{S_{oc} - S_{oa}}{S_{ac}}\right) \tag{7.2}$$

where S_{oc} is the free surface energy of oxide crystalline interface, S_{oa} is the equivalent surface term for oxide amorphous, and S_{ac} is the equivalent surface term for amorphous crystalline interfaces. Therefore, interfacial energies determine the change of crystallization temperature with thickness.

For all encapsulating materials, the increase in crystallization temperature with decreasing thickness is not universal. Simpson et al. [26] observed the crystallization of films that were just 2 nm thick which is an important observation in terms of scaling. On the other hand, as per Kissinger's analysis, activation energy (E_A) for crystallization is also observed to increase with a decrease in thickness. For example, E_A increased from 2.86 to 4.66 eV in $Ge_2Sb_2Te_5$ when the thickness of film was reduced from 20 to 5 nm [27]. The type of encapsulating layers/capping also affects the crystallization process. It is also observed that with the decrease in thickness, crystallization speed can either increase or decrease depending on the composition of phase-change material and capping layers. The melting temperature on the other hand has range from 400°C to 800°C. There are only few studies on effect of melting temperature on the reduction of film thickness, and as per those studies, conductivity and melting temperature are also reduced with a reduction in thickness of film.

With the fabrication and characterization of phase-change nanowires, material scaling in two dimensions has been studied [28]. The melting temperature has been observed to reduce with size, which will then reduce the power, and current required to amorphize the phase-change material (during RESET process). In addition, the activation energy for crystallization was noticed to get decreased in $Ge_2Sb_2Te_5$ as the size of nanowires were reduced unlike for the case of thin films. The reduction in activation energy was due to heterogeneous nucleation effects and increased surface-to-volume ratio. The non-melting amorphization is another important observation as seen in nanowires, which leads to low power consumption than melt-quenched approach. Later on, Giusca et al. [29] used CNTs as templates growing GeTe nanowires having ultra-small diameter of less than 20 nm. Such nanowires were observed to show amorphous to crystalline phase changes, which indicates the potential of phase-change devices to get scaled to ultra-small dimensions.

Scaling in three dimensions has been studied by the characterization and deposition of nanoclusters and nanoparticles of various types. Caldwell et al. [30] reported that with shrinking of nanoparticle size, crystallization temperature gets increased. Further, they predicted the least scaling limit of GeTe phase-change materials is not below 1.8 nm for proper crystallization and stability for temperature up to around 400°C. Simpson et al. [26] proposed minimum volume in $Ge_2Sb_2Te_5$ for memory operation is 1.7 nm^3. Thus, we can conclude that phase-change materials have capability to operate down to ultra-small dimensions.

The most common structure of a typical PCM cell is "mushroom" cell. In this cell, between the two metal electrodes, active phase-change material (GST) is sandwiched as shown in Figure 7.1.

Pirovano et al. [31] worked on the scaling of mushroom-type PCM cells. For isotropic scaling (assuming that the height of heater pillar, heater contact diameter, and thickness of phase-change layer all scale linearly by a factor of k where $k > 1$), the predicted behavior of parameters on scaling PCM cell is shown in Table 7.2 [18].

The scaling behavior of these parameters for the RESET (programming) of PCM mushroom cell can be explained by an ohmic representation based on simple analytical model. It assumes that close to phase-change layer/heater interface, main contribution to overall heating occurs within a hot region of diameter "D". Mathematically, the thermal resistance of heater can be written as

$$R_h^{th} = \frac{1}{\sigma_h} \frac{L_h}{\pi r_h^2} \tag{7.3}$$

where L_h is the length of heater, σ_h is the thermal conductivity of heater, and r_h is the radius of heater.

In addition, phase-change layer thermal resistance can be written mathematically as

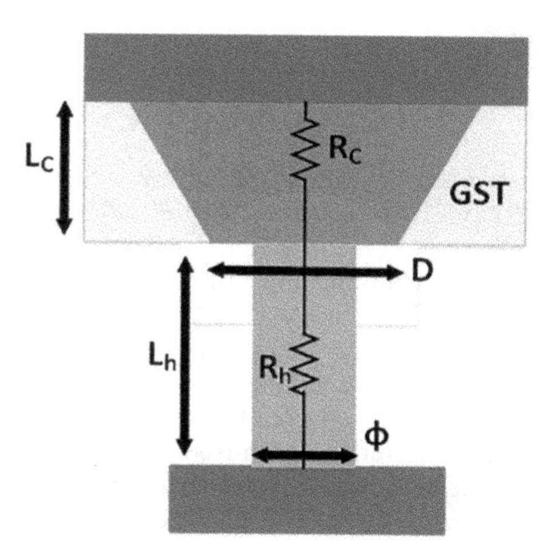

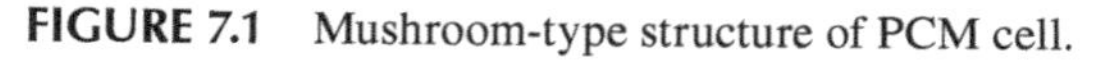
FIGURE 7.1 Mushroom-type structure of PCM cell.

TABLE 7.2
Parameter Scaling of PCM Cell

Device Parameter	Scaling Factor
Heater contact diameter (Φ)	$1/k$
Phase-change layer thickness (L_c)	$1/k$
Heater contact area	$1/k^2$
Heater height (L_h)	$1/k$
Thermal resistance (R^{th})	K
Programming current	$1/k$
Programming voltage	1
Programming power dissipation	$1/k$
SET resistance	K
RESET resistance	K
ON-state resistance (R_{ON})	K

$$R_c^{th} = \frac{1}{\sigma_c} \frac{4L_c}{\pi D(D + 2L_c)} \tag{7.4}$$

where L_c is the thickness of phase-change layer, σ_c is the thermal conductivity of phase-change layer, and D is base diameter of hot spot.

Similarly, electrical resistance of heater can be calculated as

$$R_h = \rho_h \frac{L_h}{\pi r_h^2} \tag{7.5}$$

where ρ_h is electrical resistivity of heater.

The equation for the phase-change layer resistance is given as

$$R_c = \rho_c \frac{4L_c}{\pi\phi(\phi + 2L_c)} \tag{7.6}$$

where ϕ is the diameter of heater and ρ_c is the resistivity of phase-change layer.

From Equations (7.3) to (7.6), it is observed that R_c^{th}, R_h^{th}, and R_{ON} (electrical resistance of cell during programming) will scale to k if L_c, L_h, and $\phi(=2R_h)$ are scaled isotropically by a factor of k. Thus, melting current must scale as $1/k$ if PCM cell is scaled isotropically by a factor of k. Hence, shrinking of device will reduce the melting current from 500 to 100 μA when technology node is reduced from 90 to 16 nm.

In addition, to reduce the critical operating parameters like RESET current, there is also a possibility of using alternative cell designs, cell geometry optimization, and alternative materials. One form of non-isotropic scaling in optimizations of mushroom-type cell geometry is to reduce the diameter of heater contact ϕ while keeping L_c and L_h constant. By doing this, with decreasing technology node, melting current gets decreased strongly. Besides, the thermal resistance of cell also gets altered by values chosen for L_c and L_h.

The thermal crosstalk is another important parameter to be taken into consideration while observing the effects of scaling on the PCM cell. Russo et al. [32] showed that PCM mushroom-type devices that are scaled isotropically do not have significant thermal crosstalk effect. Today, 90-nm PCM mushroom cells have been successfully fabricated with 60 nm heater contact diameter. In general, it can be stated that by using both isotropic and non-isotropic scaling approaches, challenges faced due to scaling of devices like thermal crosstalk and high programming currents can be tackled.

Around 2006, product-style PCM memories were fabricated. Later on, a variety of product-oriented PCM memories (smaller than 90 nm) such as dash type [33], μ trench [34], and pore cells [35] were developed. As compared to 400 μA RESET current in mushroom cell, μtrench cells could be programmed with a current of 400 μA at 90 nm node. Dash-type cells on the other hand were successfully designed with RESET currents below 10 μA at 20nm node. GST pore-type cells were fabricated by Wang et al. He observed extremely fast switching with 2.5 ns SET and 400 ps RESET pulses.

In addition to this, moving to horizontal and planar geometry, devices such as lateral, line, or bridge cells were fabricated to move away from the constraints put forward by the product-oriented PCM cells. Bolivar et al. [36] reported 140 μA RESET current for lateral cells having a phase-change layer thickness of 30 nm and length and width of around 90 nm. Similar values were observed by Lankhorst et al. [37] in SbTe-doped phase-change alloy having 15 nm^2 as cross-sectional area. Chen et al. [38] explained contact area dependence on RESET current, achieving RESET current of sub-100 μA values for cells of least size. Xiong et al. [39] demonstrated CNT-based PCM devices of 2–3 nm contact diameter. In addition to this, programming power and energy were also reduced. To enable cells to retain maximum heat, thermal engineering of cells needs to be done. In this context, the graphene is observed to prove as useful thermal barrier in PCM cells having about 40% lower RESET currents than the cells fabricated without graphene layer [40]. To improve operating characteristics, another method is to use different phase-change material. Phase-change super-lattice-like material is used to achieve non-melt-quenched phase switching. Using such approach, high switching speed (1ns SET pulse) and low RESET currents were reported [41]. A detailed discussion on the scaling of PCM devices can be found in Ref. [18].

7.2.3 PCM Device Architecture

The architecture of PCM device is broadly categorized into three groups [18]:

1. Self-heating device architecture
2. Built-in heater device architecture
3. Remote heater device architecture.

Self-heating device architecture is further divided into six types: (a) pillar structure, (b) self-heating wall structure, (c) line structure, (d) bridge structure, (e) via structure, and (f) metallic linear structure. The built-in heater device architecture is divided into seven groups: (a) pore structure, (b) matchstick structure, (c) ring structure, (d) μtrench structure, (e) wall structure, (f) dash-confined structure, and (g) cross-spacer structure. Similarly, remote-heater device architecture includes microheater structure.

1. **Self-Heating Device Architecture**

   In order to increase temperature and cause phase transition, self-heating device architecture is based on joule heating produced inside the chalcogenide material. It involves higher current densities, but at the same time, it has simple structure.

   a. *Pillar structure*: The compact realization of self-heating device architecture is pillar structure. For the first time, this structure was proposed with a programming current of 900 μA and diameter of 75 nm. The memory element is surrounded by electrodes on bottom and top, and is confined parallelepiped of chalcogenide material. The critical choices of this architecture are determined by patterning of phase-change material. The main limitation of this structure is that its current requirement is high.
   b. *Self-heating wall structure*: This architecture can be used to maintain control of all cell dimensions and reduce the requirement of programming current as in case of pillar structure. By the deposition of chalcogenide layer on sidewall of dielectric material, the control on the smallest dimension is achieved.
   c. *Line structure*: In this structure, thin layer strip of phase-change material is created by patterning thin layer of chalcogenide on top of dielectric structure. The structure is not optimized for applications having large density as the power dissipated in fan-out regions and the voltage drop is not sustainable in real product.
   d. *Bridge structure*: In this structure, thin chalcogenide strip crosses the top surface of dielectric layer and thus connects the two electrodes. This device was able to reach a reset voltage of 1.5 V and programming current of 90 μA, featuring a total volume of 30 mm^3 without extra losses in fan-out region of the bridge. Although this bridge structure is

more compact, it is not used for very dense array of phase cell memory cells due to its lateral extension.

e. *Via structure*: In this structure, the created hole in dielectric is filled by GST material. Unlike the pillar structure, etching of active chalcogenide material needs not to be done here.
f. *Metallic linear structure*: This structure represents the evolution of Via structure. During electrical operation, the metallic layer prevents void formation, hence improving reliability of the cell. Also, in parallel with the GST resistance, the metallic layer mitigates the drift of amorphous resistance which in turn allows better separation among partially converted (intermediate amorphous) states during the device's operating life. This enables MLC (multilevel cell) capability for PCM memory.

2. Built-in Heater Device Architecture

In order to generate heating necessary for phase transition, work has been done for introduction of dedicated element called the "heater" inside PCM cells. The properties that the heater must possess are as follows:

i. Good compatibility with phase-change alloy; that is, during operation and process with phase-change alloy, it should not have cross-contamination.
ii. In order to prevent interfacial delamination at the time of operation, it should have good adhesion with chalcogenide material.
iii. In order to have good generation of power, it should have right resistivity.
iv. It should have good endurance.
    a. *Pore structure*: Because of the ease of its realization, this structure was used to create chalcogenide-based electron device. This structure was preferred at the dawn of process integration because it does not need any chemical mechanical polishing (CMP). In this structure, the programming current of about 250 μA was reported which is competitive for similar technology node. However, it would become impractical for aggressive scaling (i.e., half pitch below 40 nm).
    b. *Matchstick structure*: After CMP became widely used, the programming current of phase-change cell in matchstick structure can be reduced with a built-in heater. To fill a hole of high aspect ratio, this structure requires a semi-metallic layer that defines critical dimension for contact area between chalcogenide alloy and matchstick. For minimizing interface resistance with phase-change material, material used needs to be clean, resilient to oxidation, and friendly for CMP. The proposals have been given to reduce programming current by engineering the interface resistance between chalcogenide and heater. In order to provide good adhesion between interfacial layers and GST, the top surface of heater is surrounded by dielectric materials laterally. Since it is difficult to find dielectric material, which will sustain current densities for an extended period of time (needed by amorphization of phase-change alloys), for such structures, the endurance is a weak spot typically. These layers can be used just as adhesion promoters in some cases.

c. *Ring structure*: In order to minimize programming current of the matchstick with little architectural difference, the ring structure was proposed. Here, only a thin layer of semi-metal or metal is deposited in order to fill hole with higher aspect ratio, and then, rest of the hole is filled with dielectric material. In this structure, the programming current is reported to get reduced. The bottom and top equivalent of thermal resistance needs to be reduced in order to effectively reduce the programming current. In addition, the dielectric material should not oxidize or damage heater material.
d. *μTrench structure*: In 180 nm technology node, the first μTrench structure was realized which was then scaled to 90 nm node. Between the heater and dielectrics surrounding it, there is always some difference in etch selectivity which makes it difficult to get a perfectly flat surface between vertical heater and chalcogenide material. This is the main issue of the μTrench structure. Nevertheless, in this structure, self-aligned evolution was shortly developed and demonstrates reset operation at this 90 nm technology node with 700 μA current.
e. *Wall structure*: This is a novel architecture, which is introduced for completing self-alignment development of the μTrench structure and for improving the control of surface chalcogenide heater. As compared to dry etching, in this structure, by CMP the top contact surface of the vertical heater is finished. This structure is quite compact, exploiting a low programming current of 200 μA at 45 nm node. Its manufacturability was demonstrated with 1.8 V as operating voltage and 1 GB PCM product. For the requirement of VLSI devices, this structure gives an optimal trade-off among various figures of merit.
f. *Dash confined structure*: In this structure, in order to control phase-change volume, replacement flow is introduced. This structure can have a minimum chalcogenide thickness of 7.5 nm and is capable of reaching programming current as low as 160 μA. In addition, for the performance of the device, the cleaning and control of recession are critical factors.
g. *Cross spacer structure*: Based on vertical heater lamina, this structure represents the extreme evolution of architecture. In this structure, on sidewalls, both the phase-change materials and heater are deposited, thus mixing the method of μTrench and self-heating wall architecture. Because of the opening for GST and heater deposition, the extension of the spacers and active sidewall dimensions begins to be quite comparable with the least dimensions that could be achieved by the already existing technology.

3. Remote Heater Device Architecture

For maintaining compact structure of device suitable for aggressive footprint of memory cell, independent optimization possibility of material parameters is offered by the heater, which is physically separated from the chalcogenide material.

a. *Microheater structure*: For inducing the phase change separated by dielectric barrier in a chalcogenide material, which stops direct electrical interaction between two parts of system, a remote heater is used. This structure is also called as microheater and can be simulated and engineered for providing exact estimation of temperature at the time of operation. In order to produce enough power for heating the chalcogenide above melting temperature (>600°C), the heater (metallic layers) requires to sustain high temperature. The possibility of re-amorphizing the crystallized volume is prevented by direct axial contact between GST and heater. Since the crystallization starts at a current surrounding the nanotube of about 25 μA, the programming current is reported to be low.

A detailed discussion on the PCM device architecture can be found in Ref. [18].

7.2.4 PCM-Based Logic Gate Design

Various numerical models have been proposed to describe the behavior of PCM [42–44]. These models of PCM can be used to implement the logic gates. In order to design gates, Reliable and ASU SPICE models [43,44] of PCM and HSPICE software have been used [45].

7.2.4.1 OR Gate Design Using PCM Logic

The circuit diagram of OR gate is shown in Figure 7.2. The OR gate consists of PCM cells arranged in parallel manner. When the binary 1, i.e., SET pulse (4 V, 300 ns), is given at both the input terminals A and B, the resistance of PCM will be set as that of crystalline phase. Thus, both the PCM cells M_1 and M_2 will conduct, and at OUTPUT, binary 1 is achieved. When at terminal A, a SET pulse is given and at terminal B a RESET pulse (6 V, 100 ns) is given, PCM cell M_1 will conduct and the cell M_2 will not conduct. Thus, at OUTPUT, binary 1is achieved. If RESET pulse is given at both the input terminals A and B, then neither of the two PCM cells will conduct. Thus, at OUTPUT, binary 0 is achieved. The simulated results of the OR gate using PCM are shown in Figure 7.3.

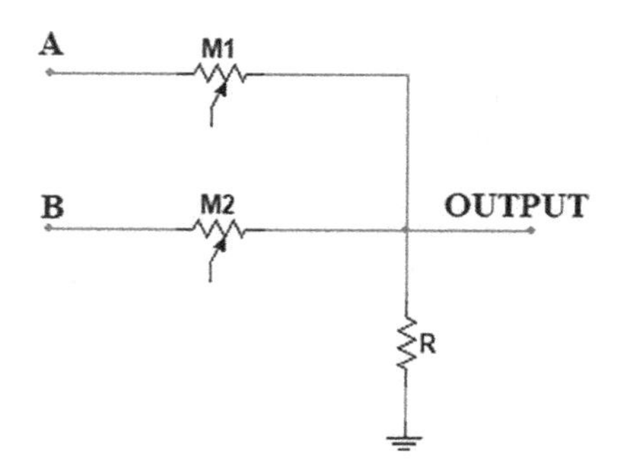

FIGURE 7.2 PCM-based circuit Diagram of OR gate.

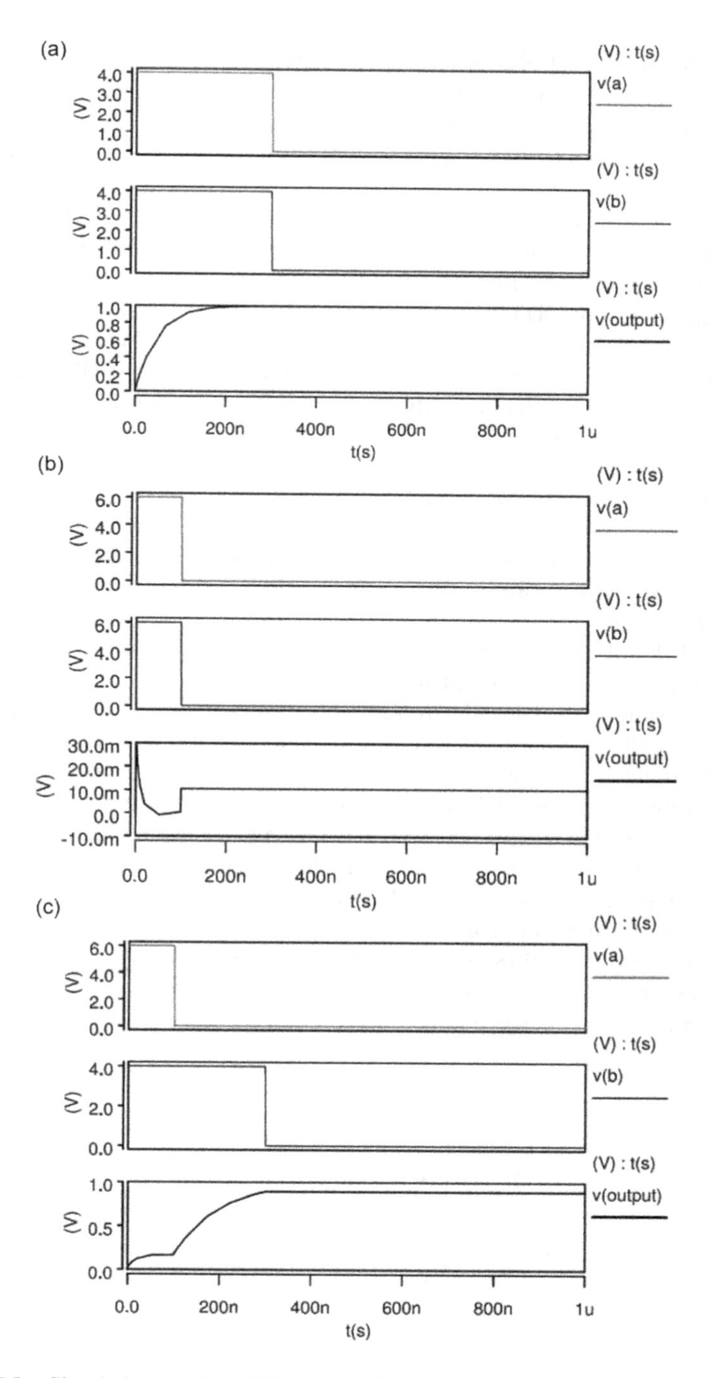

FIGURE 7.3 Simulation results of OR gate. (a) When both input terminals are in SET state. (b) When both input terminals are in RESET state. (c) When one input terminal is in SET state and the other terminal is in RESET state.

7.2.4.2 NOR Gate Design Using PCM Logic

The circuit diagram of NOR gate is shown in Figure 7.4 with $V_{DD} = 2$ V. When the binary 1, i.e., SET pulse (4 V, 300 ns) is given at both the input terminals A and B, the resistance of PCM will be set as that of crystalline phase. Hence, both PCM cells M_1 and M_2 will conduct; thus, the gate of PMOS will get 1, and transistor will not conduct. Thus, at OUTPUT, binary 0 is achieved. When at terminal A, a SET pulse is given and at terminal B a RESET pulse (6 V, 100 ns) is given, PCM cell M_1 will conduct and the cell M_2 will not conduct. The transistor again will not conduct. Thus, at OUTPUT, binary 0 is achieved. If RESET pulse is given at both the input terminals A and B, then neither of the two PCM cells will conduct and the PMOS will get 0 at its gate and will conduct. Thus, at OUTPUT, binary 1 is achieved. The simulated results of the NOR gate using PCM are shown in Figure 7.5. The range of output can be changed by varying the parameters of the circuit/PCM model.

To use the model of PCM, we require an access device at device level in order to access the PCM [46].The access device used usually is the conventional FET. The access device decides the speed, power consumption, and packing density of the chip. The selection of the access device needs to be taken into consideration in order to make design of device optimum.

The ASU PCM SPICE model [44] uses the FET as an access device to PCM and is given in Figure 7.6. The PCM needs high RESET current [47]. The FET itself needs additional power. Thus, the overall power consumption increases. The average power using FET as an access device was found to be 0.77 μW. In addition, the delay of PCM is found to be equal to 0.7505 ns using FET as an access device.

The carbon nanotube FET (CNTFET) is today used in many applications in order to increase the driving current and has replaced the FET in many areas. CNTFET requires less power than FET and has less delay. Thus, the overall power can be decreased, and the speed of the circuits using PCM model with CNTFET as access device can be increased. The average power using CNTFET as an access device was found to be 0.181 μW. In addition, the delay was found to be equal to 0.011 ns. PCM technology is one of the efficient ways of designing the logic gates because in this way the same cell can represent memory as well as logic. It can

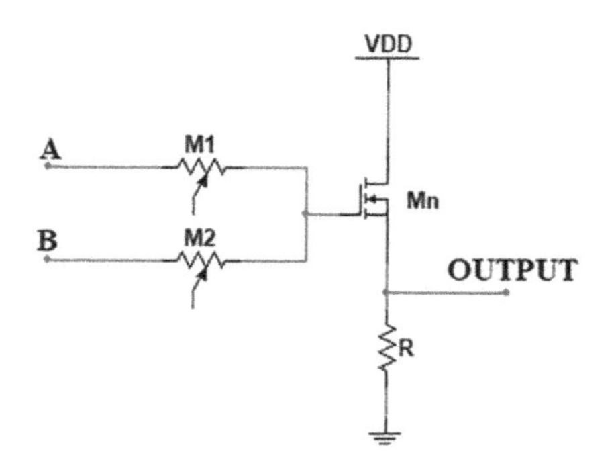

FIGURE 7.4 The circuit diagram of NOR gate using PCM-based architecture.

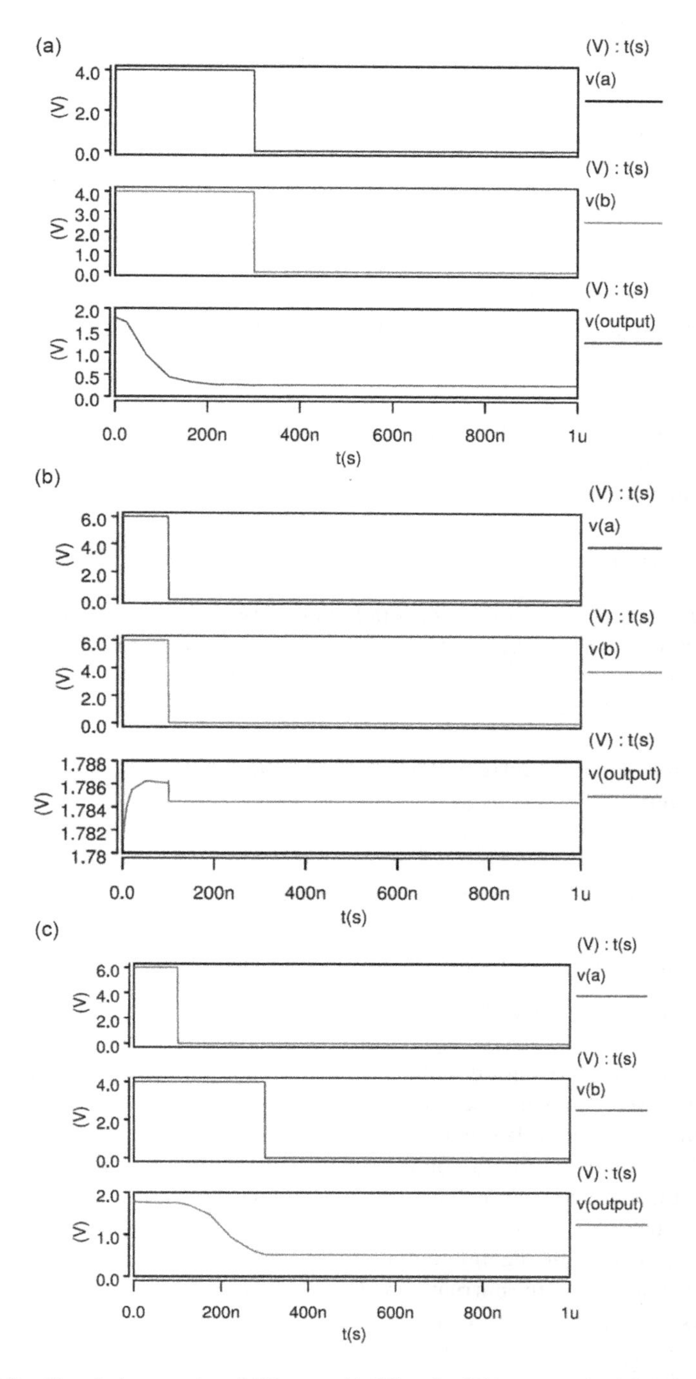

FIGURE 7.5 Simulation results of OR gate. (a) When both input terminals are in SET state. (b) When both input terminals are in RESET state. (c) When one input terminal is in SET state and the other terminal is in RESET state.

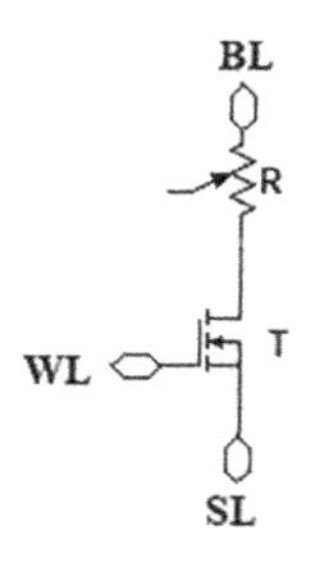

FIGURE 7.6 FET as an access device to PCM. Here, "BL" refers to bit line, "SL" refers to source, and "WL" represents word line.

easily be scaled down without any effect on the delay of circuit. Further, using CNTFET as the access device PCM model, more efficiency can be obtained vis-à-vis power and speed.

7.3 MEMRISTOR

In electrical engineering, the three fundamental two-terminal passive circuit elements are resistor (*R*), capacitor (*C*), and inductor (*L*). The four variables involved in their relationships are voltage (*V*), current (*I*), charge (*q*), and flux (φ). The relationships are the following:

i. Resistor – relating current and voltage: $v = iR$
ii. Capacitor – relating charge and voltage: $q = Cv$
iii. Inductor – relating flux and current: $\varphi = Li$

The other two relationships existing between these variables are the following:

i. Current equation – relating current and charge: $i = \dfrac{dq}{dt}$
ii. Faraday's law – relating voltage and flux: $d\varphi = Vdt$.

In total, six relationships are possible among the four variables, but only five above relationships were known. The relationship between charge and flux was missing as depicted in Figure 7.7. It was Leon Chua in 1971, who conceived the need for an additional fundamental circuit component relating charge and flux [48]. Chua proposed this missing element as "memristor," a short form for "memory resistor" as shown in Figure 7.8 [49]. A memristor is a resistor with memory and the resistance of memristor is decided by the amount and time of current flow. A memristor operates from a resistance set by the current, which flowed when it was last used and therefore remembers the last state of resistance or has memory.

Since the introduction of memristor by Chua, several researches were carried out to achieve the practical implementation of memristor. The first practical implementation of memristor was announced by Hewlett Packard (HP) laboratories in 2008 by "R Stanley Williams" [50]. This innovation led to a drastic rise of attention towards

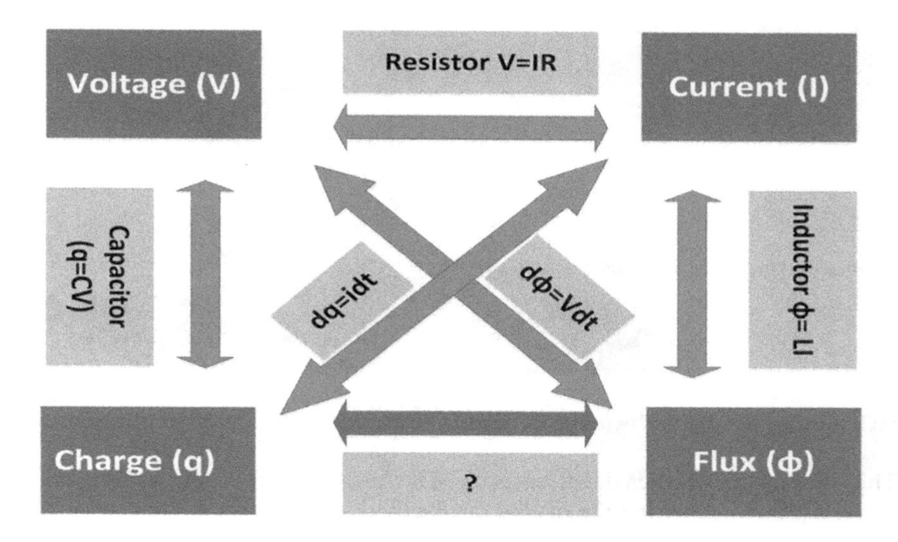

FIGURE 7.7 Relation among four fundamental circuit parameters.

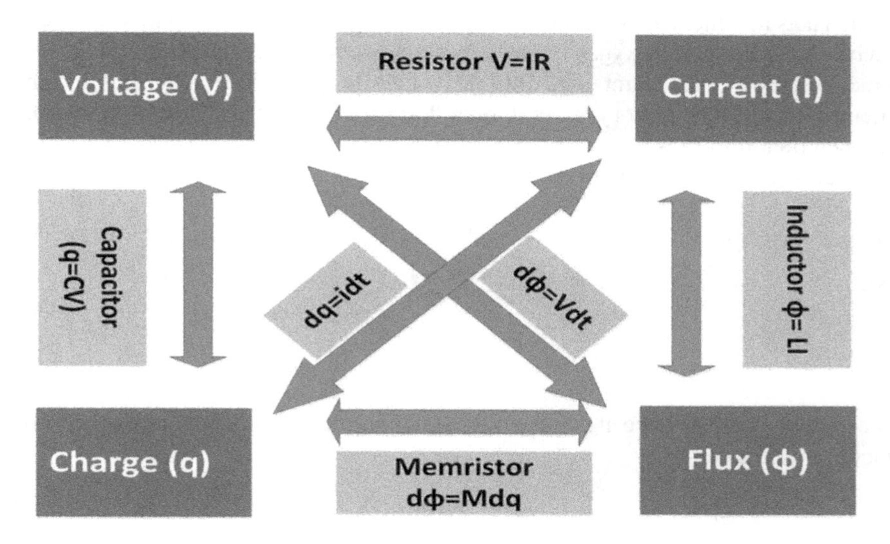

FIGURE 7.8 Relation among four fundamental circuit parameters with circuit element connecting the charge and flux.

memristor study. The memristor is constituted of a switch of two layers of titanium dioxide (TiO_2) between two conducting wires as shown in Figure 7.9. The lower TiO_2 layer has an ideal oxygen-to-titanium ratio, making it an insulator. The upper TiO_2 layer is deficient of oxygen (TiO_{2-x}) making it metallic and conductive. The oxygen deficiencies in the TiO_{2-x} act as bubbles of positive charges and are dispersed throughout the upper layer. A positive voltage on the switch repels the oxygen deficiencies in the metallic TiO_{2-x} layer and sends them into the insulating TiO_2 layer.

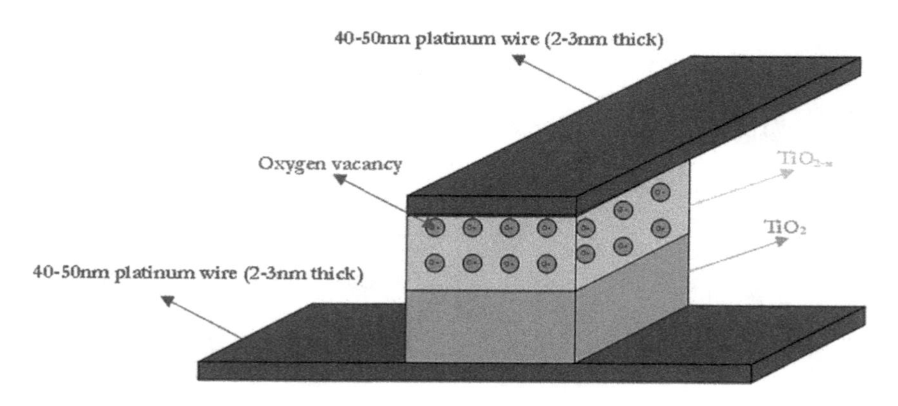

FIGURE 7.9 A switch of two layers of titanium dioxide representing a memristor.

This movement of oxygen deficiencies into TiO_2 increases the conducting percentage of TiO_{2-x}, thus increasing the overall conductivity of the entire switch. A negative voltage on the switch on the other hand pulls the positively charged oxygen bubbles out of the TiO_2, thus decreasing the overall conductivity of the entire switch. The switch acts as a memristor because when the voltage (positive or negative) across the switch is removed, the oxygen bubbles do not move from their position and the resistance of the switch is maintained until the voltage is applied [51]. After this design of memristor at HP laboratory, several other designs of memristor have been proposed in open literature [52–54].

A charge-controlled memristor is given by

$$v(t) = M(q)i(t) \tag{7.7}$$

where

$$M(q) = \frac{d\varphi(q)}{dq} \tag{7.8}$$

is called the memristance at q and has the unit of ohms (Ω), whereas a flux-controlled memristor is given by

$$i(t) = G(\varphi)v(t) \tag{7.9}$$

where

$$G(\varphi) = \frac{dq(\varphi)}{d\varphi} \tag{7.10}$$

is called the memdectance at φ and has the unit of Siemens (S).

Considering a charge-controlled memristor, the memristor constitutive relation is described analytically by a cubic polynomial:

$$\varphi = q + \frac{1}{3}q^3 \tag{7.11}$$

When a sinusoidal current source defined by

$$i(t) = \begin{cases} A\sin\omega t, & t \geq 0 \\ 0 & t < 0 \end{cases} \tag{7.12}$$

is applied across this memristor. Assuming the initial charge $q(0) = 0$, $q(t)$ can be obtained by integrating Equation (7.12):

$$q(t) = \int_0^t A\sin(\omega\tau)d\tau = \frac{A}{\omega}[1 - \cos\omega t],\ t \geq 0 \tag{7.13}$$

And using Equation (7.11), we obtain the corresponding flux given by

$$\varphi(t) = \frac{A}{\omega}(1 - \cos\omega t)\left[1 + \frac{1}{3}\left(\frac{A^2}{\omega^2}\right)(1 - \cos wt)^2\right] \tag{7.14}$$

Differentiating Equation (7.14) with respect to t, we obtain the voltage across memristor given by

$$v(t) = A\left[1 + \frac{A^2}{\omega^2}(1 - \cos\omega t)^2\right]\sin\omega t \tag{7.15}$$

The flux–charge characteristics obtained from Equations (7.12) and (7.13) are shown in Figure 7.10a. The transient responses of current and charge are shown in Figure 7.10b. Plotting the loci of $i(t)$ and $v(t)$ in the v–i plane, using Equations (7.11) and (7.15), we obtain the pinched hysteresis loop [55] shown in Figure 7.10c for unit amplitude and $\omega = 1$. This pinched hysteresis is the characteristics of a memristor and has become the identification for obtaining a memristor device. The transient responses of voltage and flux are shown in Figure 7.10d.

Pinched v–i hysteresis loop discussed above for a two-terminal device implies that the device is a memristor [55], and the pinched loop itself is useless as a model since it cannot be used to predict the voltage response to arbitrarily applied current signals, and vice versa. The only way to predict the response of the device is to derive either the ϕ–q constitutive relation, or the memristance versus state map [56].

For nonlinear devices, it is crucial to distinguish them from linear devices while applying Ohm's law. For a linear device, we have

$$V = IR \tag{7.16}$$

where R is a constant, called the resistance of the resistor, which has the unit of ohm (Ω). The resistance versus state map of a memristor also obeys Ohm's law, except that the resistance R is not a constant, but depends on a dynamical state variable x ($x = q$ in the ideal memristor case considered so far) which evolves according to a

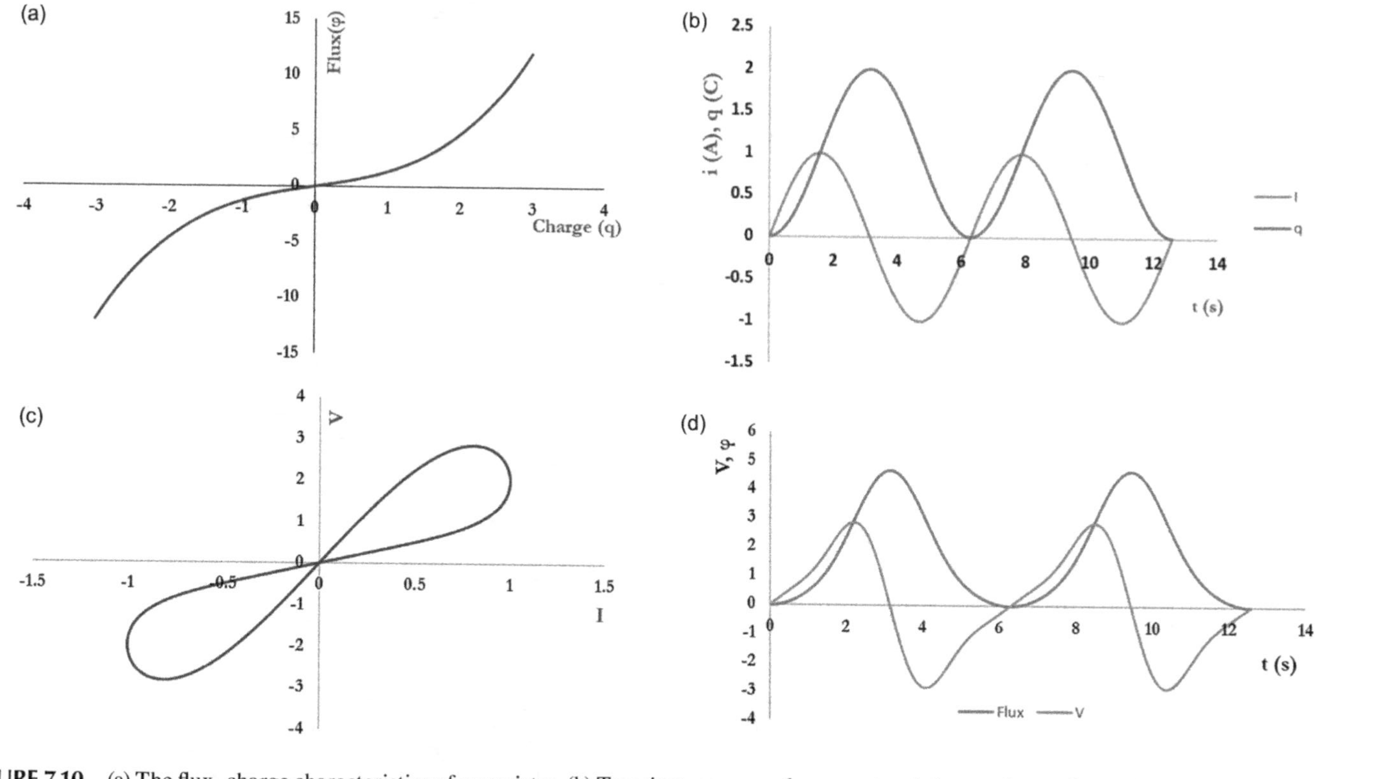

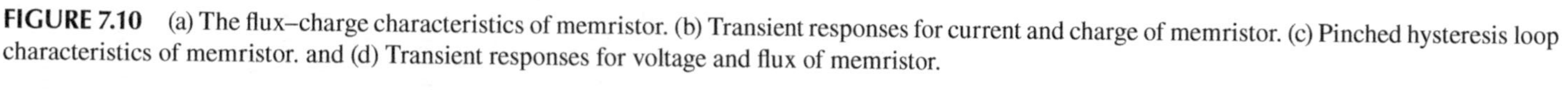

FIGURE 7.10 (a) The flux–charge characteristics of memristor. (b) Transient responses for current and charge of memristor. (c) Pinched hysteresis loop characteristics of memristor. and (d) Transient responses for voltage and flux of memristor.

prescribed ordinary differential equation, called the state equation. An ideal memristor is therefore defined by state-dependent Ohm's law:

$$V = R(x)I \tag{7.17}$$

We henceforth adopt the standard notation x to denote a state variable in mathematical system theory, where x may be a vector $x = (x_1, x_2, \ldots, x_n)$. This will be the case for many non-ideal memristors found in practice.

Memristor state equation:

$$\frac{dx}{dt} = I \tag{7.18}$$

The memristance versus state map demonstrates the complete set of small-signal memristances endowed upon a memristive device, and it is rather difficult to measure them experimentally unless the memristor can be modeled by the ideal memristor [56].

7.3.1 Memristive Devices: Switching Effects, Modeling, and Applications

Today with increase in the computational task, there is a need to process and handle a large amount of data. In such case, the von Neumann architecture becomes less significant. The biological neuromorphic system with localized storage and distributed networks becomes better choice. The emergence of novel nanoscaled devices, especially two-terminal resistive switching (RS) devices, is used to design hardware of neuromorphic system having efficient performance. One such type of device showing RS is memristor. Various causes of RS methods have been studied, and the one showing analog memristive behavior is usually used in the neuromorphic systems. The memristive devices based on valence change, electrochemical metallization (ECM), and phase-change materials are observed to show such analog memristive behavior.

In order to obtain the memristive devices, the realistic device operations are modeled by mapping the equations of memristor with the actual physical process taking place during conduction channel formation. The state variables representing memristor can be modeled by either conduction channel length method or conduction channel area method. The former method leads to non-uniform changes in conduction, whereas the later method leads to uniform changes in conductance and was first explained by Chang et al. [57,58] while explaining resistance switching in WQ_3. This method was recently used in different oxide systems also. The behavior of these oxide system devices in "neuristor" like active circuits can be predicted by memristor model-based circuit simulations.

Due to large connectivity and conductance, synaptic functions can be physically implemented in the neuromorphic circuits using memristors. Biologically, the firing pattern of pre-synaptic and post-synaptic neurons determines the weight of synapse. For efficient functioning of biological systems, several learning rules have been discovered like rate- and timing-dependent synaptic plasticity, connectivity, etc. In memristors, these necessary rules have also been successfully demonstrated. The STDP (spike timing-dependent plasticity) is an important learning rule, which states that the relative timing of pre- and post-synaptic neuron modulates the synaptic weight. The careful designing of input neuronal signals is required for the implementation of STDP with memristors. Jo et al. [59] first carried out this work with the help of CMOS/memristor circuit.

Besides STDP, rate-dependent plasticity is another learning rule. Alibart et al. [60] gave the first report on the rate dependency of memristive devices. The memristive devices were based on organic nanoparticle transistors where trapping and de-trapping of charge carriers result in synaptic plasticity. The nanoparticle transistors make scaling more difficult being a three-terminal device. Chang et al. performed systematic study on two-terminal memristor devices. In them, post-tetanic potentiation and paired pulse facilitation effects were obtained similar to those in biological systems. The devices also possess short memory retention, which is suggested by the internal decay of conduction channels. By repeated simulations, the conduction channel area is increased in the memristors which results in layer retention time and increased weight (conductance). In addition to this, Zeigler et al. [61] explained associative learning with $Ge_{0.3}Se_{0.7}$ memristors and implemented both associative and non-associative learning along with their neuron mimicking circuit.

Recent works have been done to build memristor arrays for the implementation of basic memory. With both cation and anion migration-based devices, such hybrid memristor/CMOS crossbar arrays have been demonstrated [62]. These crossbar arrays can then provide better platform for studying complex neuromorphic functions.

7.3.2 Silicon Nanowire-Based Memristive Devices

Nanoscale circuits in future have to use more efficient ways for memory storage and computation because of natural limitations of materials. In addition to charge, other fundamental state variables are phase, spin, polarity, multi-pole orientation, magnetic flux quanta, mechanical position, and other quantum states. The complete new set of possibilities for memory and logic applications is provided by physical realization of the memristor which was postulated and generalized by Leon Chua and Chua& King respectively. Under small signal, direct current, and sinusoidal excitation, a generalized model for memristive systems can be implemented. Standalone memories are targeted by one typical application: for the high-density storage, the two-terminal memristive devices are reported to have such potential [63]. The device density can be dramatically improved up to 10^{11} bits per square centimeter by high-dense crossbar arrays having memristive cross-points or by complementary logic based on two-terminal memristive devices. Moreover, to build new types of functional devices, the use of memristive effects can be exploited as new state variables with three or four terminals.

In top-down fabrication method, more advanced lithography is used which makes this approach more scalable. For the fabrication of ultra-dense nanowires that are less than 30 nm in diameter and are vertically stacked, the researchers demonstrated another solution which skips oxidation steps. The cheap solution of top-down processing applicable to both SOI and bulk-Si substrate includes the following: (1) fabrication of vertically stacked Si nanowire arrays, (2) Si nanowires with double independent gates, (3) Si nanowire with memristive functionality, and (4) ambipolar Si nanowires for memristive biosensing. The fabrication of vertically stacked Si technique is based on DRIE (deep reactive ion etching) technique that enables the structuring of Si nanowires which is not restricted by lithographic resolution. For obtaining arrays of Si nanowires (SiNWs) that are vertically stacked, this approach is utilized. Each strand can be composed of nanowires that are vertically stacked, while by the lithographic pitch, the density of horizontal strands is limited. Interconnected through Si pillars, gate-all-around field effect transistors can be built by the obtained nanowires.

In Si nanowires with double (independent)-gate technique, the device consists of crystalline SiNW which is 20 μm long attached on a SOI wafer between two Si pillars. By two independent *n++* polysilicon gates, the SiNW is then covered with this scheme: gate 1 (a main central gate) of length 7.5 μm and a gate 2 (a second gate) which is used to control the SiNW parts between the source, the main gate and drain regions, respectively [63].

The memristive devices with two terminals can be based on oxide/metal switches. These devices act as electrochemical switches whose resistance is defined by formation of metallic filament stimulated with the applied electric field polarity and related to the redox reactions in solid state. The TiO_2-based ReRAM is observed to be based on different mechanism. It is attributed to the dopant diffusion/vacancy in oxide layer. In TiO_2, the redistribution of oxygen vacancies is dependent on applied voltage polarity. It results in switching from semiconducting state to metallic state. The phase-change (PC) RAM is other type of two-terminal memristive device where the joule heating dynamics controls the phase transition between crystalline and an amorphous type controlled by voltage pulse. Another type is polymer based. By inter-posing a layer of bio-molecules, various memristive switches can be formed having properties varying from molecule-dependent switching and on molecule-independent switching [63]. Another type is based on spintronics. Pershin et al. [64] explained that electron spin polarization can be modeled as memristance that acts as a state variable and is controlled by an external voltage applied to spintronic devices. The frequency and amplitude of input signals contribute to the formation of pinched hysteresis loop in all these devices with zero crossing frequency, which is its salient feature and critical for operations of ultra-low power.

The solid-state electrolyte nanometer switch, electrochemical organic memristor, and ambipolar Si nanowire Schottky barrier FET (SiNW FET) are examples of three-terminal memristive devices. Based on the general concept of structure of FET, the three-terminal memristive devices can be classified where the memristive functionality can be inserted either by gating the memristive channel or by gate dielectric engineering. The flash memory falls in the category of capacitive memory storage of FET, in which transconductance state of channel is influenced by trap charging into gate dielectric and the volatility of charges can be tuned as per the desired response of frequency. Gated memristors are another category of which bio-memristive nanowire, electrochemical organic memristor and the solid electrolyte nanometer switch are a few examples. The conductivity change in the electrochemical organic memristor gives rise to unipolar curve of I_{ds}–V_{ds} which can be modeled as a memristor. The device in such case can be set into either diode or memristive functionality. Also, based on controlling the function formation of filament, a typical bistable resistance is shown by three-terminal solid-state electrolyte nanometer using 100 times low current as compared to current used by two-terminal operation. The ambipolar SB FET having SiNW channel lies in the group of both trap charging dielectric and gated memristor because in both gate dielectric insulator and Schottky junction, it shows dynamic trap charging mechanisms [63]. The device is similar to two back-to-back Schottky diodes [63].

For more expressive logic gates, the functionality of memristors can be viewed as state variables. By operating gate voltage polarity, memristive behavior can be tuned for SB SiNW FET. For electrons and holes, this behavior is associated with double conductance with the help of additional gate, the ambipolarity can be controlled for SB carbon nanotube so that one type of carrier conductance is blocked by it. Based on this principle, memristive SB SiNW FET having four terminals can be built [63]. Depending on the behavior of controlling signal applied at the channel of Si wire, two operating modes are reported as (a) voltage-controlled four-terminal memristive devices and (b) current-controlled four-terminal memristive devices. With the help of dual-gate configuration, a four-terminal voltage-controlled memristive Schottky barrier SiNW FET is produced so that a portion of channel is controlled by one of the two gates between the main gate and drain/source contacts. In order to control ambipolarity imbalance, this configuration is exploited [63]. On the other hand, instead of V_{ds}, the current I_{ds} is required for obtaining current-controlled type of four-terminal memristive SiNW Schottky barrier FET [63]. The V_{gs} is then compared with the output voltage. The resulted hysteresis can then be used as latch, whose location can be adjusted by using bias current of different value in the V_{out}-V_{in} plane.

There are various applications of memristive devices. Some of them are listed below:

1. *Field programmable gate array*: In past few years, many novel field programmable gate array building blocks and architectures have been proposed. For instance, ReRAMs-based routing structures have shown tremendous potential. To route signals via less resistive paths, for switchboxes, a cross-point using non-volatile ReRAM as switches in FPGA has been proposed. For optimization of time, ReRAM switch-based routing elements were exploited. Moreover, in FPGA, ReRAM-based MUX was incorporated into the routing structure [63].
2. *Neuromorphic circuits*: In the neuromorphic circuits, the memristive devices are used to form artificial synapses following the Hebbian rule of learning which is based on spike rate-dependent plasticity. It is interesting to note down that for emulating the biological system behavior, circuits based on memristive devices can be built. One example is memristive-based energy efficient integrate and neuron circuit that uses the ReRAM bistability to model both refractory period and short time spike event. In addition,

the weighted connections of the model of perceptron can be emulated by using analog programmability of ReRAM devices [63].

3. *Multi-valued logic*: In order to replace CMOS-based traditional Boolean logic, MVL (multi-valued logic) circuit design technique is used. Recently for the design of GAA SB SiNW transistors that are vertically stacked showing ambipolar nature, a fully compatible process has been demonstrated. The gates exploiting denser logic functions can be designed by using the concept of ambipolarity [63].
4. *Current and temperature sensor*: Based of Schottky barrier SiNW transistors, pA current and temperature-sensing device can be fabricated having very low power consumption, Besides, for heterogeneous integration, process flow compatibility makes these devices suitable for both logic and sensing applications [63].

7.3.3 Memristor-Based Logic Design

Classical digital computation is performed on binary logic, which has only two possible values (0, 1) or (true, false). MVL design and computation can be used as an alternative to traditional digital design and computation [65]. MVL implementation can reduce chip area and power consumption by more than 50% [66]. MVL modules have already found place in binary logic integrated circuits for better performance in CMOS technologies [67]. For MVL systems, there exists the optimal base number system R such that as R increases, there is an increase in the number of information per connection in digital signal processing [68]. For economical implementation of a digital signal processor, it has been analytically proven that ternary system provides the best solution [69]. With an increase in radix R, the tolerance of system decreases, whereas the complexity of system increases [69]. Therefore, ternary system seems natural extension of binary system. The ternary logic design is believed to achieve simplicity and energy efficiency in digital circuits, as ternary logic reduces the complexity of interconnects and increases chip density. The ternary number system is represented in two forms: balanced presented as (–1, 0, 1) and unbalanced presented as (1, 0, 2) [70].

Ternary logic systems have been proposed to improve the functionality of digital design [71,72]. Other works include implementation, simulation, and fabrication of ternary logic design using standard ternary inverter (STI) in CMOS technology [71] and a similar STI based on CNTFET technology [66]. The design based on STI CMOS technology uses resistors to achieve ternary logic [71]. The resistor circuit consumes more power and area on a chip. CNTFET provides low power design, but it has compatibility issue with available CMOS technology. Memristor-based ternary circuits are also proposed for ternary logic design [73]. Memristor is suited for realization of ternary logic systems as it can handle multiple states without using any extra hardware. Ternary logic gates are designed using CMOS technology and memristor [73]. It has been proposed that practical memristor [51] is compatible with CMOS; therefore, it is optimal for design [74]. In addition, memristors are relatively smaller in size, which provides better fabrication density [75]. The reconfigurable and storage property of memristor can help us to design advanced computer architectures [75]. It has been proposed that multiple logic operations can be performed on data that is stored on the memristor device itself [56].

The ternary logic has one level more than the binary logic. These three levels can be either balanced or unbalanced. In balanced mode –1, 0 and 1 are used to represent false, undefined, and true logic states of ternary logic. The basic operations of ternary logic with Y as output and X_i and X_j as input can be defined as follows:

$$Y_{\text{inverter}} = -X \tag{7.19}$$

$$Y_{\text{OR}} = \max(X_i, X_j) \tag{7.20}$$

$$Y_{\text{AND}} = \min(X_i, X_j) \tag{7.21}$$

$$Y_{\text{NOR}} = \overline{\max(X_i, X_j)} \tag{7.22}$$

$$Y_{\text{NAND}} = \overline{\min(X_i, X_j)} \tag{7.23}$$

In digital circuits, different logic levels are represented using different voltage levels. In the presented designs, three voltages –1.2, 0, and 1.2 V have been used to represent low, intermediate, and high logic levels, respectively. If we consider *L*, *I*, and *H* as low, intermediate, and high logic levels of ternary logic design, then the truth table of ternary inverter is given in Table 7.3, whereas the truth table of ternary AND, OR, NAND, and NOR gates is given in Table 7.4.

Using the above equations and truth table, the logic circuits have been designed.

The circuit diagrams of inverter, NAND, and NOR gates are shown in Figure 7.11a–c, respectively. Since the resistance of memristor varies with respect to applied input voltage, the memristors connected with gate terminal help to get ternary logic. The input voltage is applied to memristor and has three different levels to get different logic states of ternary logic. The resistance value across memristor also depends on how these terminals are connected. As shown in Figure 7.11b, c, the NAND and NOR circuits are different from each other only the way memristor is connected.

The inverter is implemented using two MOS transistors (M_1 and M_2) and one memristor. If the applied input logic is low, then transistor M_2 is turned on, and M_1 is turned off. As a result, no current flows in the CMOS inverter and output voltage follows V_{ss}. If the input logic becomes high, then transistor M_2 turns off

TABLE 7.3
Truth Table of Ternary NOT Gate

Input	Output
L	H
I	I
H	L

TABLE 7.4
Truth Table of Different Logic Gates in Ternary Logic System

Input Vin1	Input Vin2	Output AND	Output OR	Output NAND	Output NOR
L	L	L	L	H	H
I	L	L	I	H	I
H	L	L	H	H	L
L	I	L	I	H	I
I	I	I	I	I	I
H	I	I	H	I	L
L	H	L	H	H	L
I	H	I	H	I	L
H	H	H	H	L	L

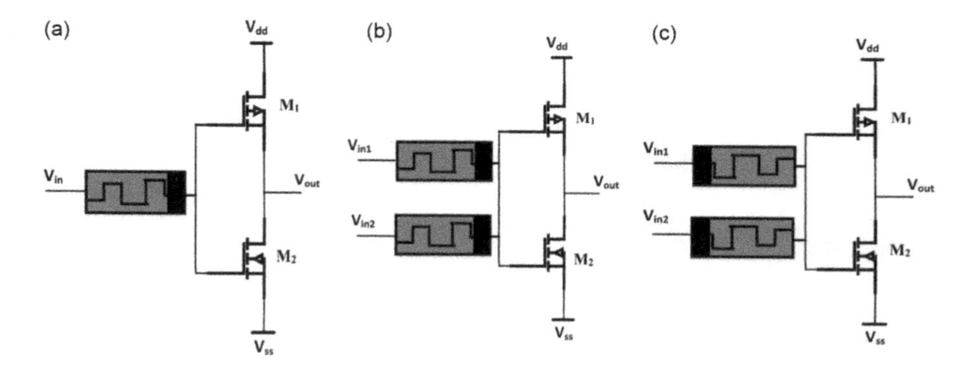

FIGURE 7.11 Circuit diagram of ternary: (a) Inverter, (b) NAND gate, and (c) NOR gate.

and M_1 is on. Again, no current flows in the CMOS transistor, but this time output voltage follows V_{dd}. Finally, if the applied input logic is intermediate, then both the transistors will turn on and output voltage will depend on the drop across transistors M_1 and M_2. The NAND and NOR gates have two input voltages connected with two memristors. The output of these two memristors is provided as input to CMOS inverter circuit. The output voltage of memristors determines the state of ternary logic. The voltage produced by memristors depends on input voltage and follows the logic as described above. For three different input voltage levels, memristors generate different outputs, which determine the ternary logic implementation of these gates.

The operation of the gates was examined with standard 180 nm CMOS and OTA-based memristor emulator, and the simulations were carried out in Cadence Virtuoso software [76]. The supply voltages were chosen as $V_{DD} = +1.2\,\text{V}$ and $V_{SS} = -1.2\,\text{V}$. The voltage levels for ternary input are −1.2, 0, and 1.2 V. The simulation results of inverter, NAND gate, and NOR gate are shown in Figure 7.12a–c, respectively.

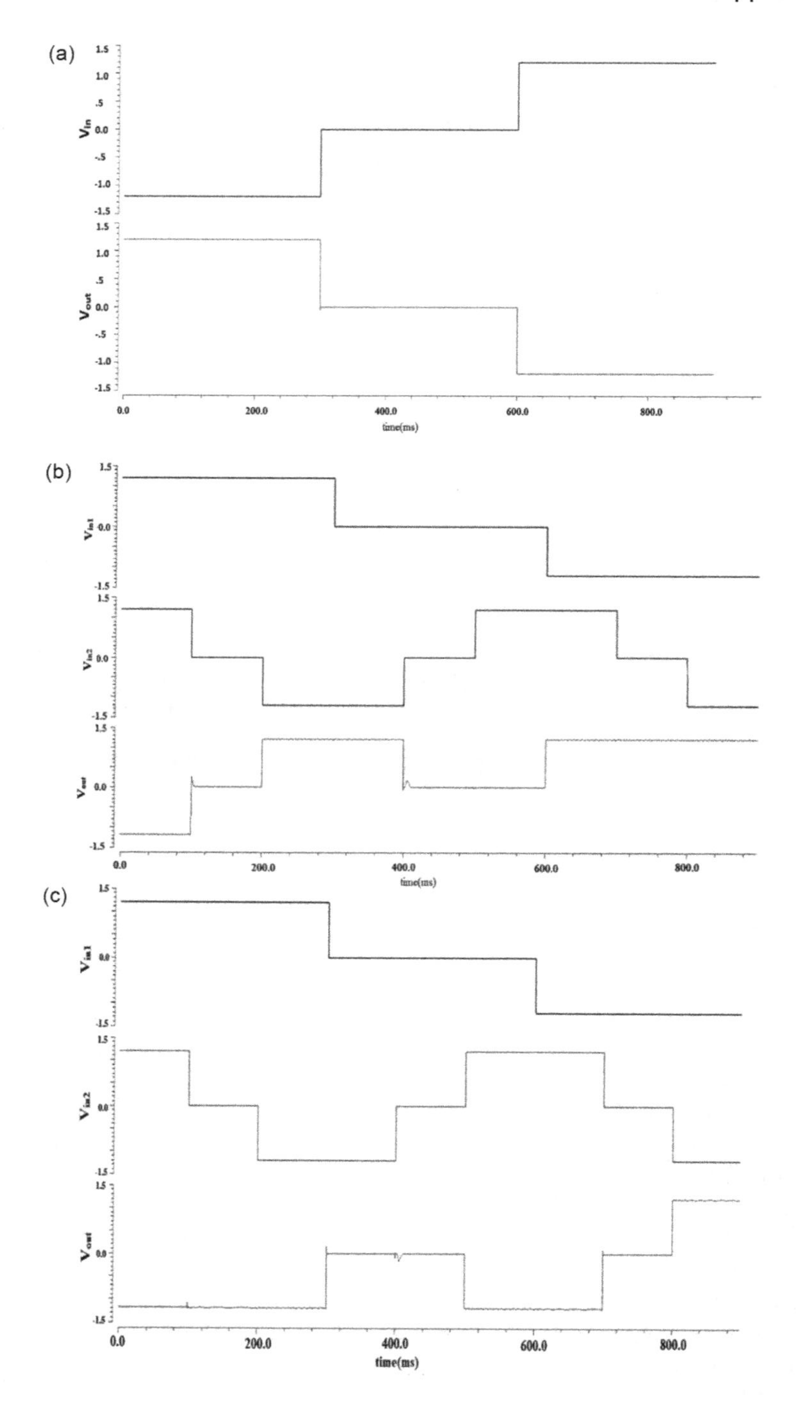

FIGURE 7.12 Simulation result of (a) ternary inverter, (b) ternary NAND gate, and (c) ternary NOR gate.

TABLE 7.5
Truth Table of 2:4 Ternary Decoder

A	B	C (carry)	S (sum)
L	L	L	L
I	L	L	I
H	L	L	H
L	I	L	I
I	I	I	I
H	I	I	I
L	H	L	H
I	H	I	I
H	H	H	L

TABLE 7.6
Truth Table of Ternary Half Adder

A_1	A_0	D_3	D_2	D_1	D_0
L	L	L	L	L	H
I	L	L	I	L	I
H	L	L	H	L	L
L	I	L	L	I	I
I	I	I	I	I	I
H	I	I	I	L	L
L	H	L	L	H	L
I	H	I	L	I	L
H	H	H	L	L	L

The presented gates can be used to implement higher-order ternary circuits. The truth tables of 2:4 ternary decoder and ternary half adder are given in Tables 7.5 and 7.6, respectively. The simulation results of 2:4 ternary decoder and ternary half adder are given in Figure 7.13a, b, respectively.

7.4 RESISTIVE RANDOM-ACCESS MEMORY (RRAM)

7.4.1 Physical Structure of RRAM

It has been observed that in some insulators, the change of resistance occurs under the application of the applied electric field. This property of change of resistance has recently been investigated for developing future non-volatile memories [77]. In literature, a number of different phenomena have been studied for this resistive transition. Among these phenomena, an overview of those considered for non-volatile memory applications is given in Figure 7.14. Apart from these phenomena, the redox-based nanoionic chemical effects, which include ECM effects, valence change

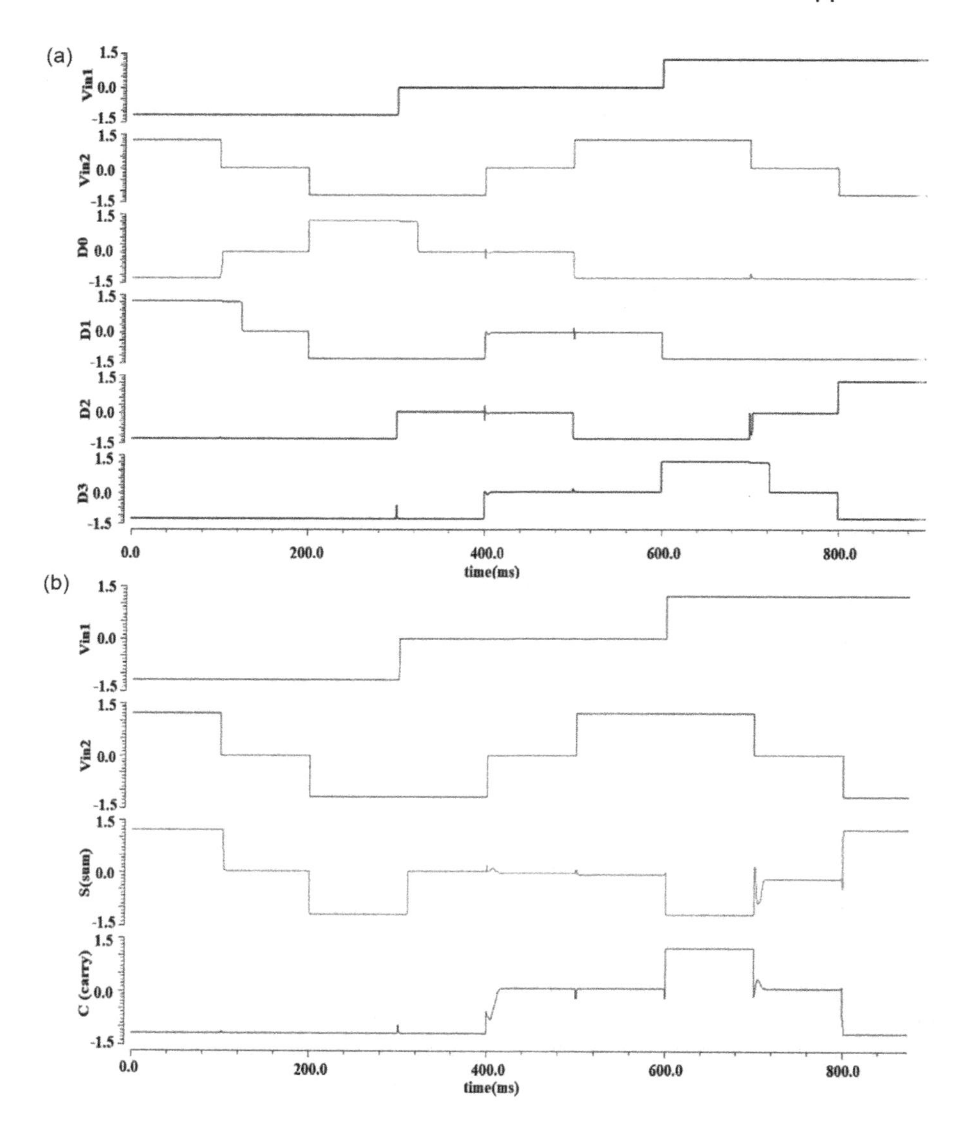

FIGURE 7.13 Simulation result of (a) 2:4 ternary decoder and (b) ternary half adder.

memory (VCM) effects, and thermochemical memory (TCM) effects, have been studied in great detail.

An RRAM consists of a RS memory cell having a metal–insulator–metal structure referred to as MIM structure. The structure is also referred to as metal-oxide-metal (MOM) structure. It comprises an insulating layer (I) sandwiched between the two metal (M) electrodes. The schematic view of a RRAM cell is shown in Figure 7.15.

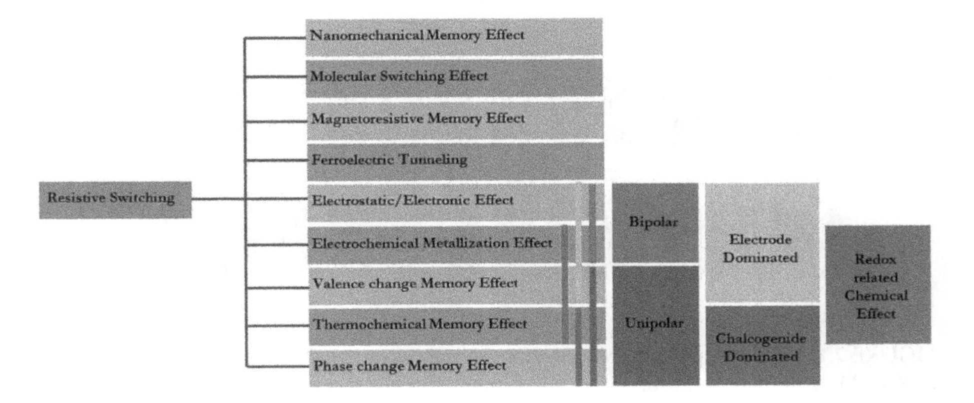

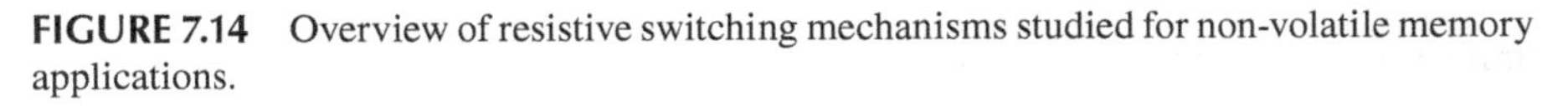

FIGURE 7.14 Overview of resistive switching mechanisms studied for non-volatile memory applications.

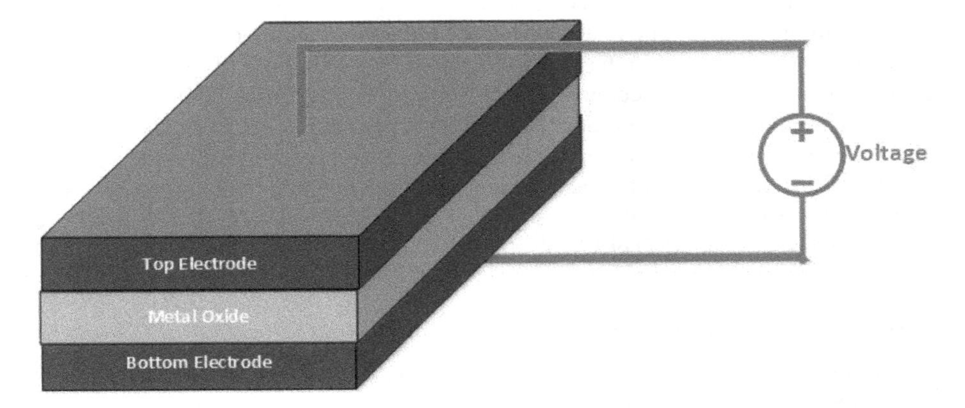

FIGURE 7.15 Schematic of metal–insulator–metal structure of RRAM [4].

7.4.2 Resistance Switching Materials

The resistance-switching phenomenon has been observed in a variety of oxides, but binary metal oxides have been extensively studied as preferred switching materials for future non-volatile memory applications primarily due to their compatibility with the CMOS BEOL processing. Various metal-oxide-based materials exhibiting the non-volatile resistance switching such as hafnium oxide (HfOx) [78], titanium oxide (TiOx) [79], tantalum oxide (TaOx) [80], nickel oxide (NiO) [81], zinc oxide (ZnO) [82], zinc titanate (Zn_2TiO_4) [83], manganese oxide (MnOx) [84], magnesium oxide (MgO) [85], aluminum oxide (AlOx) [86], and zirconium dioxide (ZrO_2) [87] have drawn the most attention and have been studied extensively during the past several years. These metal oxides are deposited usually by pulse laser deposition (PLD), atomic layer deposition (ALD), and reactive sputtering. However, ALD is a widely preferred method owing to its ability to precisely control the thickness and

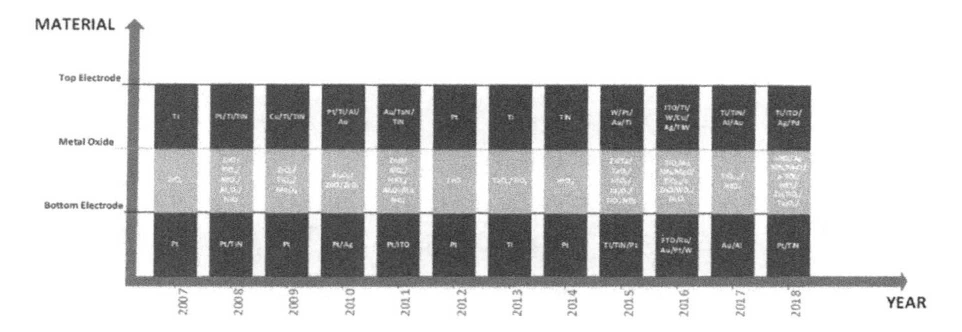

FIGURE 7.16 The graphical representation for the list of materials used for the fabrication of RRAM up to the year 2018.

uniformity of the thin film [88]. The graphical representation for the list of metal oxide materials that have been used recently in the fabrication of the RRAM device along with various combinations of materials used for the top and bottom electrodes is shown in Figure 7.16. The material choice for the fabrication of RRAM gives it an edge as MOM structures can be easily fabricated by making use of oxides currently used in the semiconductor technology. The bottom electrode material in RRAM usually is platinum, which is a bit hard to etch. For single device structure, RRAM can share the same bottom electrode, whereas, for the crossbar architecture, the separate bottom electrodes are used for each device. They can be obtained by physical vapor deposition (PVD) and lift off successively. The top electrode and the RS layer are deposited using either ALD or PVD.

7.4.3 Resistance Switching Modes

The application of the external voltage pulse across the RRAM cell enables a transition of the device from a high resistance state (HRS) or OFF state generally referred to as logic value "0" to a low resistance state (LRS) or ON state generally referred to as logic value "1" and vice versa. The RS phenomenon is considered to be the reason behind this change of resistance values in a RRAM cell. An RRAM is initially in the HR state. To switch the device from the HRS to the LRS, the application of the high voltage pulse enables the formation of conductive paths in the switching layer and the RRAM cell is switched into a LR state. This process which occurs due to the soft breakdown of the MIM structure is usually referred to as "electroforming," and the voltage at which this process occurs is referred to as forming voltage (V_f). The forming voltage depends on the cell area and oxide thickness [89]. Now, to switch the RRAM cell from the LRS to HRS, the voltage pulse referred to as the RESET voltage (Vreset) is applied which enables this switching transition, and the process is referred to as the "RESET" process [90]. The HRS of the RRAM can be changed to LRS on the application of the voltage pulse. The voltage at which the transition occurs from HRS to LRS is referred to as SET voltage (*V*set), and the process is referred to as the "SET" process [91]. To efficiently read data from RRAM cell, a small read voltage which will not disturb the current state of the cell is applied to

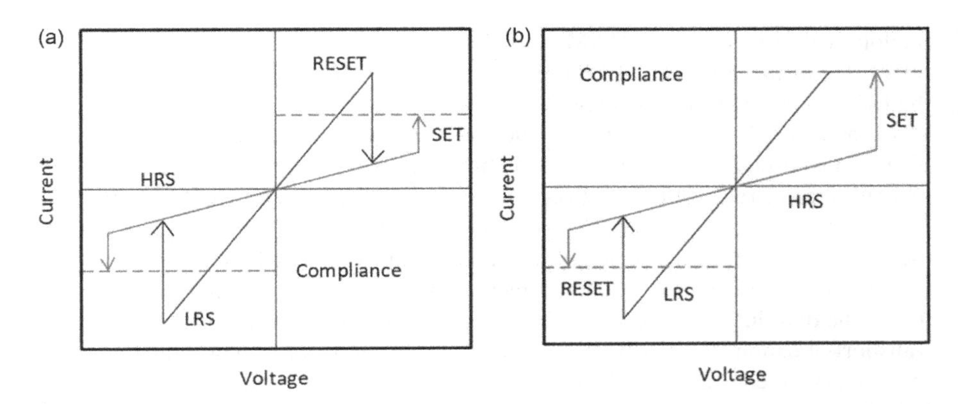

FIGURE 7.17 I–V curves for RRAM: (a) Unipolar switching and (b) bipolar switching.

determine whether the cell is in the Logic 0 (HRS) or the Logic 1 (LRS) state. Since both LRS and HRS retain their respective values even after the removal of applied voltage, RRAM is a non-volatile memory. Depending on the polarity of the applied voltage, the RRAM can be classified into two types of switching modes: (1) unipolar switching and (2) bipolar switching [92]. In unipolar switching, the switching (set and reset processes) of the device between various resistance states does not depend on the polarity of the applied voltage; that is, switching can occur on applying a voltage of the same polarity but different magnitude as shown in Figure 7.17a. In bipolar switching, the switching (set and reset process) of the device between various resistance states depends on the polarity of the applied voltage; that is, a transition from a HRS to LRS occurs at one polarity (either positive or negative), and the opposite polarity switches the RRAM cell back into the HRS as depicted in Figure 7.17b. In unipolar switching, joule heating is interpreted as the physical mechanism responsible to rupture a conducting filament during reset operation. In bipolar switching, on the other hand, the migration of charged species is the main driving force for CF dissolution although joule heating still contributes to accelerate the migration. In order to ensure there is no permanent breakdown of the dielectric switching layer during the forming/set process of RRAM, a compliance current (Icc) is enforced for the RRAM device. The compliance current (Icc) is usually ensured by a cell selection device (transistor, diode, and resistor) or by a semiconductor parameter analyzer during the off-chip testing.

7.4.4 Resistive Switching Mechanism

The switching of the RRAM cell is based on the growth of CF inside a dielectric. The CF is a channel having a very less diameter of the order of nanometers, which connects the top and the bottom electrodes of the memory cell. A LRS with high conductivity is obtained when the filament is connected and the high resistance state (HRS) results when the filament is disconnected with a gap between the electrodes [68]. Based on the composition of the CF, RRAM can be classified into the following two types: (1) metal ion-based RRAM also referred to as conductive bridge random-access memory (CBRAM) and (2) oxygen vacancies filament-based RRAM referred to as the "OxRRAM." It must be noted here that CBRAM is also known as the ECM memory, whereas "OxRRAM" is also known as VCM.

In metal ion-based RRAM also referred to as "CBRAM," the physical mechanism that is responsible for resistive switching is based on the migration of metal ions and subsequent reduction/oxidation (redox) reactions [93]. The CBRAM structure consists of an oxidizable top electrode (anode) such as Ag, Cu, and Ni; a relatively inert bottom electrode (cathode), e.g., W and Pt; and a sandwiched metal oxide layer between the two electrodes. The filament formation in such memory cells occurs due to the dissolution of the active metal electrodes (most commonly Ag or Cu), the transport of cations (Cu^+ or Ag^+), and their subsequent deposition or reduction at the inert bottom electrode [94]. Thus, the resistive switching behavior of this type of RRAM is dominated by the formation and dissolution of the metal filaments.

To obtain a better understanding of the switching mechanism of metal ion-based CBRAM, an example of Ag/a-ZnO/Pt RRAM cells is considered [95]. A general schematic illustration depicting the switching process of CBRAM cell is shown in Figure 7.18. The pristine state of the CBRAM memory cell is depicted in Figure 7.18a. The Ag top electrode is an active component in the filament formation, whereas the bottom Pt electrode is inert. On the application of the positive voltage bias to the Ag top electrode, the oxidation ($Ag \rightarrow Ag^+ + e^-$) occurs at the top electrode because of which Ag^+ cations are generated and get deposited into the dielectric layer (a-ZnO) from the Ag electrode. The negative bias on the Pt bottom electrode attracts the Ag^+ cations and as such the reduction reaction ($Ag^+ e^- \rightarrow Ag$) occurs at the bottom electrode. Thus, the Ag^+ cations are reduced to Ag atoms and accumulate until the conducting bridge is formed (Figure 7.18b–d), and the RRAM device is said to exhibit LRS. This process is referred to as the "SET." When the polarity of the applied voltage is reversed, the highly conducting filament dissolves almost completely and the device is said to be in the HRS. This process is referred to as "RESET" and is depicted in Figure 7.18e. In oxygen vacancy-based RRAM (OxRRAM), the physical mechanism that is responsible for resistive switching is similar to the one discussed for memristors. A comparison of OxRRAM with CBRAM based on various operational parameters reveals the striking difference

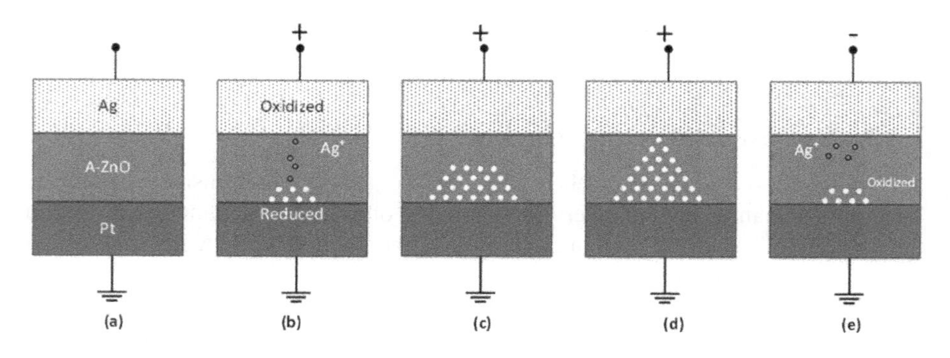

FIGURE 7.18 Schematic of the switching mechanism of CBRAM [95].

in terms of the endurance characteristics of these RRAM memory types. This dissimilarity is because the conducting filaments of CBRAM are composed mainly of metal atoms which are relatively easier to drift and diffuse compared to oxygen vacancies, thus causing the degradation of the retention time and endurance characteristics of CBRAM compared to the OxRRAM. Although the switching mechanism of the both RRAM types is different, there are many common characteristics between the two of them. The only significant difference is that endurance for OxRRAM is significantly higher than that for CBRAM.

7.4.5 Performance Metrics of Resistive Random-Access Memory (RRAM)

7.4.5.1 Write Operation

In write operation, the important factors to be considered are the voltage amplitude (V_{wr}) and the length of the write pulse (t_{wr}). In order to be compatible with the current VLSI technology, the range for voltage must be restricted within a few hundred millivolts and few volts. The smallest write voltage has been reported to be 0.6 V in a CBRAM cell [96]. As far as pulse widths are concerned, the desired length should be below 100 ns to compete with DRAM and win over flash. If the pulse widths for write operations are reduced below 10 ns, then RRAM has the chance to achieve the efficiency close to high-performance SRAM. Researchers have reported subnanosecond pulses capable of introducing resistive transitions adding to the possibility of RRAM replacing SRAM [31].

7.4.5.2 Read Operation

As in case of write operation, read operations are also characterized by the amplitude of the voltage required and length of pulse, but an additional constraint here is the requirement of a minimum read current (I_{rd}). The read voltages (V_{rd}) should be significantly smaller than V_{wr} in order to prevent any change of the resistance during the read operation. Because of constraints enforced by circuit design, V_{rd} cannot be less than approximately one-tenth of V_{wr} [96]. Taking into consideration the prowess of current sense amplifiers, the amount of I_{rd} should be preferably larger than 1 μA to allow for fast detection. As far as pulse widths in read cycles are concerned, they should either be of the same duration as the write pulse or smaller. As far as RRAM is concerned, V_{rd} as small as 0.1 V has been already demonstrated [96].

7.4.5.3 Resistance Ratio

The ratio between the maximum achievable resistance to that of the minimum achievable resistance is known as resistance ratio. In practice, resistance ratios as small as 1.2–1.3 have been utilized in MRAM designs, but researchers suggest that in order to comply with the current design strategies, a ratio of greater than 10 would allow for small and very efficient sense amplifiers. Small and efficient sense amplifier designs would make RRAM cost-effective against flash memory. Resistance ratios are where RRAMs usually are far ahead of their counterparts, with ratios as high as 10^8 already reported [97].

7.4.5.4 Endurance

The maximum number of write cycles before a device starts to show performance deterioration is endurance. Traditional flash-based memory designs can usually have endurance rating between 10^3 and 10^7. In order to be competitive enough with current flash designs, RRAM should provide the same endurance as flash or better. Current research trends and data collected over the years show that more than 10^{12} write cycles can be extrapolated from RRAM devices with 10^{12} endurance cycles already demonstrated [96].

7.4.5.5 Retention

The maximum amount of time a device can retain its memory state after being programmed is a measure of retention. From a universal non-volatile memory point of view, the required retention time should be greater than 10 years while the device is kept under a thermal stress of up to 85°C and a constant stream of read pulses. Data and research over the years support the fact that RRAMs can retain data for a period greater than 10 years [96].

7.4.5.6 Uniformity

In RRAM cell, poor uniformity of various device characteristics is one of the significant factors limiting the manufacturing on a wider scale. The switching voltages, as well as both the HRS and the LRS resistances, are among the parameters exhibiting a high degree of variation. The variations of the resistance switching include temporal fluctuations (cycle-to-cycle) and spatial fluctuations (device-to-device). The stochastic nature of the formation and rupture of CF is believed to be the main reason for these variations. Cycle-to-cycle and device-to-device variability is a major hindrance for information storage in RRAM devices [98]. The observation of cycle-to-cycle variability is influenced by the change in the number of oxygen vacancy defects that arise in the CF due to its stochastic nature of formation and rupture during the switching event [99]. Due to this random nature of the CF, the prediction and the precise control of the shape of the CF becomes extremely challenging. This variability becomes worse as the compliance limit (i.e., compliance current "Icc") is reduced. A lot of research has been conducted to improve the uniformity of RRAM, and several methods have been explored for the same. One of the methods utilizes the concept of inserting nanocrystal seeds which confine the paths of the CF by enhancing the effect of local electric field [100]. In addition to the materials engineering approach, a novel programming method has also been suggested to reduce fluctuations. A multi-step forming technique was implemented in W/HfO_2/Zr/TiN [101]-based RRAM to minimize the overshoot current due to the parasitic effects. A multi-step forming technique results in the gradual formation of the filament; thus, a low set/reset current is achieved improving the switching characteristics of the device. Various other methods such as constant voltage forming and hot forming (usually referred to as forming at a higher temperature) have also been investigated to effectively reduce the resistance variations [102]. Another method of achieving high uniformity is by applying a pulse train rather than a single pulse to a RRAM cell [103]. This approach not only results in improved uniformity but also enhances the multilevel capability of a RRAM cell.

7.4.5.7 Effect of Operating Temperature and Random Telegraph Noise

To achieve a reliable performance of the RRAM device, the effect of operating temperature and random telegraph noise (RTN) should be minimum. It is observed that the resistance of both the LRS and HRS states undergoes variations because of the change of operating temperature. The on–off resistance ratio (Ron/Roff) also decreases with an increase in temperature affecting the memory performance. RTN is another factor that affects the performance of RRAM. RTN decreases the memory margin between the HRS and LRS because of the extensive fluctuations in the read current during the read operation. Due to the effect of RTN, the read margin, scaling potential, and the multilevel cell capability of a RRAM cell are greatly affected [104]. The RTN is attributed to the trapping and de-trapping of electrons in the proximity of the CF in LRS, whereas it occurs in the tunneling gap in the HRS state. It is observed that with the decrease in operation current, the amplitude of RTN increases, thus highly affecting the HRS level. Therefore, it is necessary to ensure the additional resistance margin to achieve reliable performance.

7.4.6 RRAM-Based Non-Volatile Memory (NVM) Design

With the realization that resistive switching cells can be used as building blocks for non-volatile memory, the research interest in RRAM has increased exponentially in recent years. The pinched hysteresis loop in their current–voltage relationships qualifies them as memristive systems [49], which further extends their use beyond just non-volatile storage elements. Studies in recent years have been conducted to explore the possibility of using RRAM behaviors for neuromorphic and stochastic computing domains. From a system designer's perspective, in order to explore these possibilities, it is important to be able to model the behavior of RRAM cells. Computing models allow the systems to be simulated in a cheap and efficient manner. It allows newer systems to be tested without the need for actual hardware. Accurate models are important in understanding device behaviors and optimize designs. RRAM technology is still in its early stages; a proper and accurate model will prove to be an efficient tool for studying designs and standardize implementations. A considerable research effort in recent times has been focused towards RRAM modeling. Several models have been given for both bipolar and unipolar RRAMs. The models for bipolar RRAM include Chua's model [48], Simon's tunnel barrier model [105], Yakopcic's model [106], threshold adaptive memristor model (TEAM) [107], voltage threshold adaptive memristor model (VTEAM) [108], Stanford/ASU model [109], physical electro-thermal model [110], Huang's physical model [111], Bocquet's bipolar model [112], the Berco–Tseng model [113], and the Gonzalez-Cordero et al. bipolar model [114]. The models for unipolar RRAM include random circuit breaker model [115], filament dissolution model [116], and Window function models [117].

One of the most commonly used RRAM array architectures is the one-transistor-one-resistor (1T1R) architecture [78,87,118,119] shown in Figure 7.19, wherein a RRAM device is connected in series with the access device usually a MOSFET. This memory cell (row i and column j) can be utilized as a part of a greater memory array, where M_{ij} represents the RRAM cell, and X_{ij} is the CNTFET device. WL_i is the word

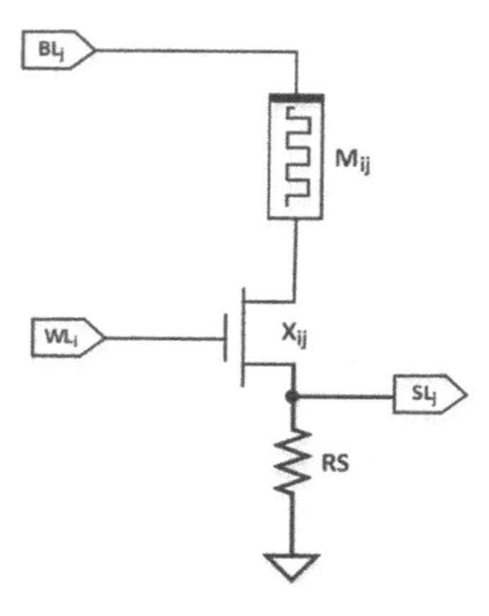

FIGURE 7.19 Schematic of memory cell having RRAM in series with access device (MOSFET).

line signal selector of row i, BL_j is the bit line signal selector of column j, and SL_j is the sensing line of column j that connects to the sense amplifier. Replacement of the MOSFET by a CNTFET not only helps in enhancing low power profile of the RRAM memory cell but can potentially increase the operating speed at well [118]. The 1T1R architecture with CNTFET as access device is shown in Figure 7.20.

A typical 1T1R configuration is developed in HSPICE with RRAM cell connected in series with an n-type CNTFET device. Both the SET and the RESET pulse widths in this design are fixed (10 ns). To initiate SET process, i.e., to make the RRAM cell transition to LRS from its initial HRS, the word line WLi is turned on, and the SET pulse is applied on the bit line BL_j. For the case of RESET process, i.e., to make the RRAM cell transition from HRS to LRS, the word line WLi is turned

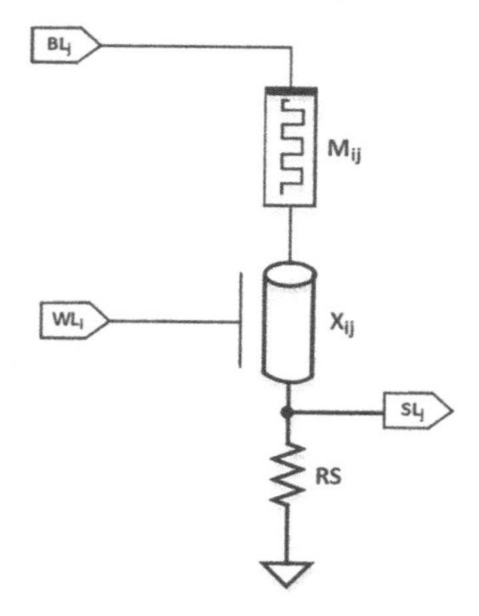

FIGURE 7.20 Schematic of memory cell having RRAM in series with CNTFET access device.

on, the RESET pulse is applied on the bit line BL_j. Note that for both the procedures of SET and the RESET, the corresponding signal is obtained on sense line SLj which is connected to the ground through RS. In Figure 7.21a, we plot the input voltage WL_i as a function of time. On WL_i, a 3 V signal with rise and fall times $t_r = t_f = 0.1$ ns is applied. For SET or RESET process, the operating voltage is applied through BL_j. The applied BL_j signal is plotted in Figure7.21b. To initiate a SET process (logic 1), a voltage pulse $V_{set} = 6$ V is applied on BL_j with a pulse duration of 40 ns and $t_r = t_f = 0.1$ ns. The CF is thus formed in the device, and the RRAM cell switches to LRS from the initial HRS. For the RESET process (logic 0), a RESET pulse $Vreset = -6$ V, with pulse duration of 40 ns and $t_r = t_f = 0.1$ns is applied on BL_j. The applied pulse leads to the dissolution of CF resulting in HRS. To read the contents of the RRAM cell, a voltage pulse $V_{Read} = 3$ V is applied on BL_j. Figure7.21c shows the sense line signal of the memory cell. The resistance chosen for the design is RS = 4.5 kΩ. It must be noted that in the SL, the negative voltages are because of the internal capacitances of RRAM and CNTFET. The BL_j signal of the memory cell (Figure7.21b) shows a series of set-read-read-reset-read-read operations performed on memory cell, and at the sense line signal SL_j obtained is depicted in Figure 7.21c. A comparison of power consumption of Figures 7.19 and 7.20 is given in Table 7.7. As is evident from the simulated observations, the ITIR RRAM using CNTFET as an access device consumes less power as compared to ITIR RRAM using MOSFET as a access device.

Figure 7.20 can be extended for the design of higher order memory architecture as shown in Figure 7.22 for 3 × 3 memory array. The access to an particular cell M_{21} needs a high signal to be applied at WL_2, and low voltages are required for word lines WL_1 and WL_3. The components of column 1 are affected by the control signal at BL_1. The voltage levels in RS_1 are affected due to the resistance variation

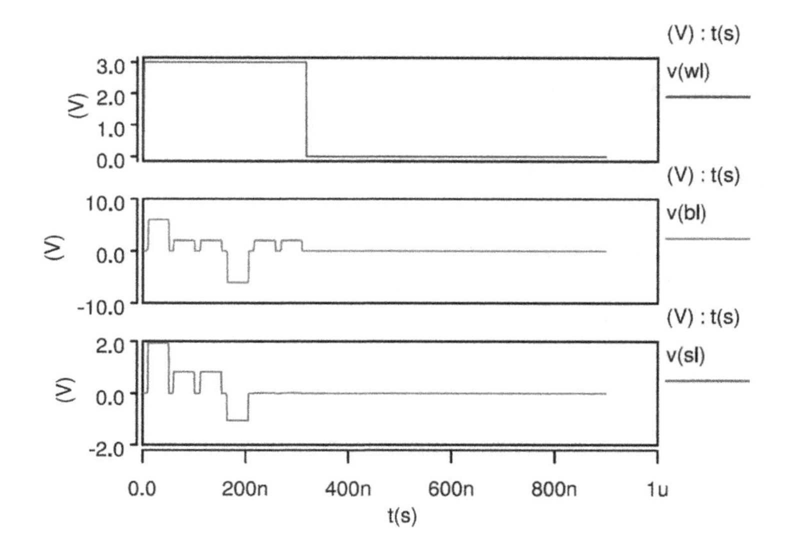

FIGURE 7.21 Voltage signals plotted as a function of time. (a) Word line signal applied to gate terminal of n-type CNTFET. (b) Bit line signal applied to top electrode of RRAM. (c) Sense line voltage for the RRAM CNTFET model.

TABLE 7.7
Comparison of Power Consumption

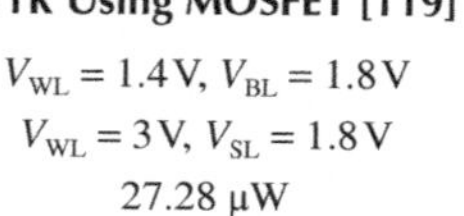

Parameter	1T1R Using MOSFET [119]	1T1R Using CNTFET [118]
SET voltage	$V_{WL} = 1.4\,V$, $V_{BL} = 1.8\,V$	$V_{WL} = 3\,V$, $V_{BL} = 6\,V$
RESET voltage	$V_{WL} = 3\,V$, $V_{SL} = 1.8\,V$	$V_{WL} = 3\,V$, $V_{SL} = -6\,V$
Power consumption	27.28 μW	6.99 μW

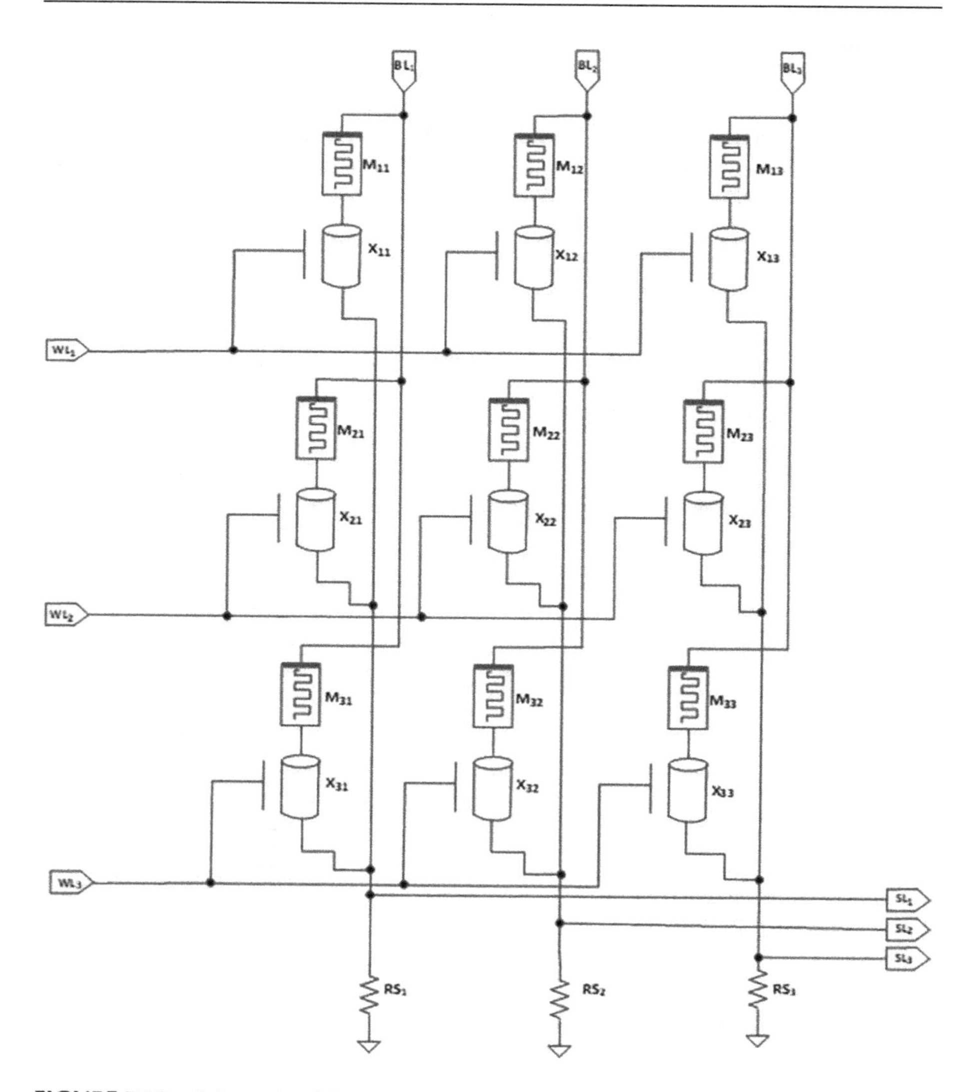

FIGURE 7.22 Schematic of 3 × 3 memory array based on ITIR structure with RRAM and CNTFET.

at M_{21}. Similarly, access to M_{33} requires WL_3 to be high and low voltages on WL_1 and WL_2, respectively. The BL_3 control signal influences the elements of column 3, while change in resistance at M_{33} affects the signal amplitude levels of RS_3.

Figure 7.23 shows the sequential set-read-read-reset-read-read operation for the memory cells (M_{11}, M_{12}, M_{13}...M_{31}, M_{32}, M_{33}). Figure 7.23a represents the word line sequence (WL_1, WL_2, WL_3) for the memory array. Figure 7.23b shows the bit line-sequence (BL_1, BL_2, BL_3) for the memory array, while shows sense line signal (SL_1, SL_2, SL_3) is shown in Figure 7.23c.

The availability of a range of resistive states means that instead of storing binary values, RRAM devices can be used to store more than two values in the form of different resistive states. In literature, a number of material systems have been used for the study of multilevel capability in RRAMs [120]. While both the SET and RESET regimes can be used for multilevel storage by controlling the set current and reset voltage, respectively, it is easier to control voltage inputs using the access device making reset regimes more favorable [120]. In using the RESET transition, two different approaches can be taken to program the device to a desired resistive state: either at the beginning of each programming cycle, the device is first programmed to LRS using a strong SET pulse followed by a partial reset by applying a weak RESET pulse or by programming the desired state from a prior reset state [121]. The former scheme while being easier to implement will have negative effects on the endurance, but the latter usually requires the previous state to be known. Both these operations are usually carried out by controlling the gate voltage at the access device. The application of a controlled RESET allows the partial dissociation of the CF grown during the SET operation.

The instruction codes and the data are transferred by making use of buses between various units in a computer system having von Neumann architecture because of the separate computing and memory unit. This data transferring process results in increased energy consumption and time delay, which is commonly referred to as "von Neumann bottleneck." For reducing the impact of von Neumann bottleneck [122], computing based on the crossbar RRAM array is suggested which alters the memory and computing operations in the same core. In addition, to obtain high integration density and low cost [123], the highly compact two-terminal device structure and $4F^2$ array architecture of the RRAM are beneficial. For example, to obtain simple Boolean logic functions such as "logic NOT," "logic AND," and "logic OR," we require multiple transistors and each single transistor takes 8–10 F^2 area. The same logic functions can be realized by making use of two or three RRAM cells, resulting in the total area of around 10 F^2 area only [124]. Till date, several methods have been suggested for realizing Boolean logic functions [125]. Boolean computing is significantly more established compared to other non-Boolean computing paradigms such as neuromorphic computing and quantum computing. Therefore, energy and cost efficiency of CPU or MCU can be enhanced without the need to develop new algorithms or software, although there is still a lack of technical solution on how to implement complex computing tasks in a crossbar array. Thus, most of the research to date focuses on only basic level logic demonstration as it is quite complex to use RRAM array as a complete computing unit.

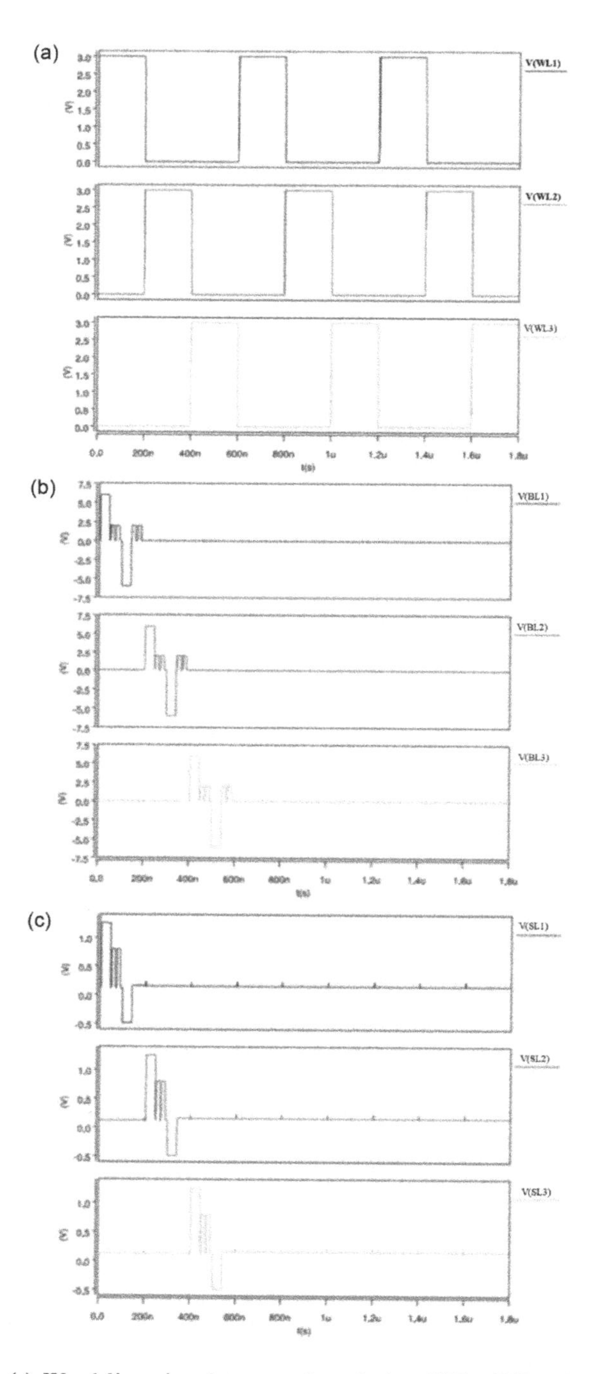

FIGURE 7.23 (a) Word line signal temporal evolution (WL_1, WL_2, WL_3), (b) bit line signal temporal evolution (BL_1, BL_2, BL_3), and (c) sense line signal temporal evolution (SL_1, SL_2, SL_3).

The other applications of RRAM include neuromorphic computing. Compared to CMOS-based neuromorphic network, neuromorphic computing based on RRAM array offers advantages in terms of on-chip weight storage, online training, and scaling up to much larger array size [126]. In addition, the processing speed of RRAM improves by three orders of magnitude, whereas the power consumption rate is reduced by four orders of magnitude [127].

7.5 SUMMARY

In the semiconductor industry, any electronic device with only two electrical terminals is usually referred to as a non-volatile resistance-switching memory device if the device can exhibit one of two resistance values over a sufficiently long time period, without consuming any power, and can be switched from a low-resistance state to a high-resistance state and vice versa, by applying either a short voltage pulse or a short current pulse, of appropriate amplitude and polarity, across the two device terminals, and such that the resistance state at any time, either low or high, can be sensed by applying a relatively much smaller sensing voltage pulse or current pulse of some preset waveform, across the same terminals. By uncovering the physical operating mechanisms taking place internal to the device, one could construct a model that exhibits these two memristances. Some devices of this behavior, which have been discussed in this chapter, are PCM), memristor, and RRAM. The devices present a great scope in wide domains ranging from non-volatile logic/memory design to neuromorphic computing to security applications. While the devices promise some untouched applications in the electronics industry, they also pose some serious challenges vis-à-vis their implementation and quality metrics. Therefore, a continuous research effort is going on to perfect their implementation and exploit their real-life industrial applications.

REFERENCES

1. J.S. Meena, S.M. Sze, U. Chand, T.Y. Tseng, *Nanoscale Res. Lett.* 9(1), 526 (2014).
2. F. Pan, S. Gao, C. Chen, C. Song, F. Zeng, *Mater. Sci. Eng. R Rep.* 83, 1–59 (2014).
3. H.S.P. Wong, S. Salahuddin, *Nat. Nanotechnol.* 10(3), 191, (2015).
4. S. Yu, *Resistive Random Access Memory (RRAM) From Devices to Array Architectures*, Morgan and Claypool Publishers (San Rafael, CA, 2016).
5. H.S.P. Wong, S. Raoux, S. Kim, J. Liang, J.P. Reifenberg, B. Rajendran, M. Asheghi, K.E. Goodson, *Proc. IEEE* 98(12), 2201–2227 (2010).
6. G.W. Burr, M.J. Breitwisch, M. Franceschini, D. Garetto, K. Gopalakrishnan, B. Jackson, B. Kurdi, C. Lam, L.A. Lastras, A. Padilla, B. Rajendran, S. Raoux, R.S. Shenoy, *J. Vac. Sci. Technol. B Nanotechnol. Microelectron. Mater. Process. Measurement, and Phenom.* 28(2), 223–262 (2010).
7. M. Wuttig, N. Yamada, *Nat. Mater.* 6(11), 824 (2007).
8. D. Apalkov, B. Dieny, J.M. Slaughter, *Proc. IEEE* 104(10), 1796–1830 (2016).
9. B. Dieny, R.C. Sousa, J. Herault, C. Papusoi, G. Prenat, U. Ebels, D. Houssameddine, B. Rodmacq, S. Auffret, L.D.B. Prejbeanu, M.C. Cyrille, B. Delaet, O. Redon, C. Ducruet, J.P. Nozieres, I.L. Prejbeanu, *Int. J. Nanotechnol.* 7(4–8), 591–614 (2010).
10. C. Chappert, A. Fert, F.N. Van Dau, *Nat Mater.* 6(11), 813–823 (2007).

11. R. Tetzlaff, *Memristors and Memristive Systems*, Springer (New York, Heidelberg, Dordrecht, London, 2013).
12. L.I.Y. Tao, L.O.N.G. ShiBing, L.I.U. Qi, L.U. HangBing, L.I.U. Su, L.I.U. Ming, *Chin. Sci. Bull.* 56(28–29), 3072 (2011).
13. L. Zhu, J. Zhou, Z. Guo, Z. Sun, *J. Materiomics* 1(4), 285–295 (2015).
14. D. Ielmini, *Semicond. Sci. Technol.* 31(6), 063002 (2016).
15. P. Lorenzi, R. Rao, F. Irrera, *J. Vac. Sci. Technol. B Nanotechnol. Microelectron. Mater. Process. Meas. Phenom.* 33(1), 01A107 (2015).
16. C. Ye, J. Wu, G. He, J. Zhang, T. Deng, P. He, H. Wang, *J. Mater. Sci. Technol.* 32(1), 1–11 (2016).
17. X. Yang, *J. Nanosci.* 2016, 8132701 (2016).
18. A. Redaelli, *Phase Change Memory: Device Physics, Reliability and Applications*, Springer International Publishing AG (Cham, 2018).
19. S.R. Ovshinsky, "Symmetrical current controlling device", U.S. Patent 3 271 591, 1966.
20. S. Kim, B. Lee, M. Asheghi, G.A.M. Hurkx, J.P. Reifenberg, K.E. Goodson, H.S.P. Wong, "Thermal disturbance and its impact on reliability of phase-change memory studied by Micro-thermal stage", *2010 IEEE International Reliability Physics Symposium (IRPS), Anaheim, CA*, 2010.
21. S.R. Ovshinsky, *Phys. Rev. Lett.* 21, 1450–1453 (1968).
22. R. Fallica, J.L. Battaglia, S. Cocco, C. Monguzzi, A. Teren, C. Wiemer, E. Varesi, R. Cecchini, A. Gotti, M. Fanciulli, *J. Chem. Eng. Data* 54(6), 1698–1701 (2009).
23. S.R. Ovshinsky, *Phys. Rev. Lett.* 21(20), 1450–1453 (1968)
24. M.H. Cohen, H. Fritzsche, S.R. Ovshinsky, *Phys. Rev. Lett.* 22, 1065–1068 (1969).
25. M. Fowler, "*Electrons in one dimension: the Peierls Transition*", 2007.
26. R.E. Simpson, M. Krbal, P. Fons, A.V. Kolobov, J. Tominaga, T. Uruga, H. Tanida, *Nano Lett.* 10(2), 414–419 (2009).
27. W. Xiaoqian, L. Shi, T. Chong, Z. Rong, H. Koon, *Jpn. J. Appl. Phys.* 46(4B), 2211 (2007).
28. S.H. Lee, Y. Jung, R. Agarwal, *Nat. Nanotechnol.* 2(10), 626–630 (2007).
29. C.E. Giusca, V. Stolojan, J. Sloan, F. B€orrnert, H. Shiozawa, K. Sader, M.H. Rümmeli, B. Büchner, S.R.P. Silva, *Nano Lett.* 13(9), 4020–4027 (2013).
30. M.A. Caldwell, S. Raoux, R.Y. Wang, H.S.P. Wong, D.J. Milliron, *J. Mater. Chem.* 20 (7), 1285–1291 (2010).
31. A. Pirovano, A.L. Lacaita, A. Benvenuti, F. Pellizzer, S. Hudgens, R. Bez, "Scaling analysis of phase-change memory technology," in *IEEE International Electron Devices Meeting, 2003 (IEDM'03 Technical Digest)*, pp. 29–6, 2003.
32. U. Russo, D. Ielmini, A. Redaelli, A.L. Lacaita, *IEEE Trans. Electron Devices* 55(2), 515–522 (2008).
33. D.H. Im, J.I. Lee, S.L. Cho, H.G. An, D.H. Kim, I.S. Kim, H. Park, D.H. Ahn, H. Horii, S.O. Park, U.I. Chung, "A unified 7.5 nm dash-type confined cell for high performance PRAM device," in *2008 I.E. International Electron Devices Meeting*, 1–4, 2008.
34. F. Pellizzer, A. Benvenuti, B. Gleixner, Y. Kim, B. Johnson, M. Magistretti, T. Marangon, A. Pirovano, R. Bez, G. Atwood, "A 90nm Phase Change Memory Technology for Stand-Alone NVM Applications", in *Proceedings Symposium on VLSI Technology*, 2006.
35. W.J. Wang, L.P. Shi, R. Zhao, K.G. Lim, H.K. Lee, T.C. Chong, Y.H. Wu, *Appl. Phys. Lett.* 93(4), 043121 (2008).
36. P.H. Bolivar, F. Merget, D.H. Kim, B. Hadam, H. Kurz, Inst. Semicond. Electron. RWTH Aachen University Sommerfeldstr 24, 52056 (2004).
37. M.H. Lankhorst, B.W. Ketelaars, R.A.M. Wolters, *Nat. Mater.* 4(4), 347–352 (2005).
38. Y.C. Chen, C.T. Rettner, S. Raoux, G.W. Burr, S.H. Chen, R.M. Shelby, M. Salinga, W.P. Risk, T.D. Happ, G.M. McClelland, M. Breitwisch, "Ultra-thin phase-change bridge memory device using GeSb", in *International Electron Devices Meeting*, pp. 777–780, 2006.

39. F. Xiong, A.D. Liao, D. Estrada, E. Pop, *Science* 332(6029), 568–570 (2011).
40. C. Ahn, S.W. Fong, Y. Kim, S. Lee, A. Sood, C.M. Neumann, M. Asheghi, K.E. Goodson, E. Pop, H.S.P. Wong, *Nano Lett.* 15(10), 6809–6814 (2015).
41. D. Loke, L. Shi, W. Wang, R. Zhao, H. Yang, L.T. Ng, K.G. Lim, T.C. Chong, Y.C. Yeo, *Nanotechnology* 22(25), 254019 (2011).
42. C. Dao-Lin, S. Zhi-Tang, L. Xi, C. Hou-Peng, C. Xiao-Gang, *Chin. Phys. Lett.* 28(1), 018501 (2011).
43. D.Biolek, M.D. Ventra, Y.V. Pershin, *Radioengineering* 22(4) (2013).
44. Z. Xu, K.B. Sutaria, C. Yang, C. Chakrabarti, Y. Cao, "Hierarchical modeling of phase change memory for reliable design", *IEEE International Conference on Computer Design*, pp. 115–120, 2012.
45. S. Mehraj, F.A. Khanday, "PCM based Logic Design and Performance analysis using CNTFET as Access Device", *3rd International Conference on Communication and Electronics Systems (ICCES 2018)*, 15–16 October 2018, PPG Institute of Technology, Coimbatore, India, pp. 568–571, 2018.
46. K. Gopalakrishnan et al., Highly-scalable novel access device based on Mixed Ionic Electronic conduction (MIEC) materials for high density phase change memory (PCM) arrays", *Symposium on VLSI Technology (VLSIT)*, 2010. DOI: 10.1109/VLSIT.2010.5556229.
47. G. Atwood, R. Bez, "90nm Phase Change Technology with μTrench and Lance Cell Elements. In VLSI Technology", *International Symposium on Systems and Applications*, 2007, VLSI-TSA 2007, pp. 1–2, 2007.
48. L.O. Chua, *IEEE Trans. Circuit Theory* 18, 507–519 (1971).
49. L. Chua, S.M. Kang, *Proc. IEEE* 64(2), 209–223 (1976). DOI: 10.1109/PROC.1976.10092
50. D.B. Strukov, G.S. Snider, D.R. Stewart, R.S. Williams, *Nature* 453, 80–83 (2008).
51. W. Stanley, *IEEE Spectrum* 45, 28–35 (2009). DOI: 10.1109/MSPEC.2008.4687366.
52. S.H. Jo, K.H. Kim, W. Lu, *Nano Lett.* 9, 870–874 (2009).
53. M.J. Lee, C.B. Lee, D. Lee, S.R. Lee, M. Chang, J.H. Hur, Y.B. Kim, C.J. Kim, D.H. Seo, S. Seo, U.I. Chung, I.K. Yoo, K. Kim, *Nat. Mater.* 10, 625–630 (2011).
54. Y. Yang, P. Sheridan, W. Lu, *Appl. Phys. Lett.* 100, 203112 (2012).
55. L. Chua, "If It's Pinched It's a Memristor," in *Memristors and Memristive Systems*, R. Tetzlaff Eds. Springer (New York, 2014).
56. L. Chua, "Resistance switching memories are memristors," in *Memristor Networks*, A. Adamatzky, L. Chua Eds. Springer (Berlin, 2014).
57. T. Chang, S.H. Jo, K.H. Kim, P. Sheridan, S. Gaba, W. Lu, *Appl. Phys. A Mater. Sci. Process.* 102, 857–863 (2011).
58. T. Chang, S.-H. Jo, W. Lu, *ACS Nano* 5, 7669–7676 (2011).
59. S.H. Jo et al., *Nano Lett.* 10(4), 1297–1301 (2010).
60. F. Alibart, S. Pleutin, D. Guerin, C. Novembre, S. Lenfant, K. Lmimouni, C. Gamrat, D. Vuillaume, *Adv. Funct. Mater.* 20, 330–337 (2010).
61. M. Ziegler, R. Soni, T. Patelczyk, M. Ignatov, T. Bartsch, P. Meuffelsand, H. Kohlstedt, *Adv. Funct. Mater.* 22, 2744–2749 (2012).
62. K.H. Kim, S. Gaba, D.Wheeler, J.M. Cruz-Albrecht, T. Hussain, N. Srinivasa, W. Lu, *Nano Lett.* 12, 389–395 (2012).
63. R. Tetzlaff Ed, *Memristors and Memristive Systems*, Springer (New York, 2014).
64. Y.V. Pershin, M. Di Ventra, *Phys. Rev. B* 78(11), 113309 (2008).
65. M. Mukaidono, *IEEE Trans. Comput.* C-35(2), 179–183 (1986).
66. S. Lin, Y. B. Kim, F. Lombardi, *IEEE Trans. Nanotechnol.* 10, 217–225 (2012).
67. D.A. Rich, *IEEE Trans. Comput.* 35(2), 99–106, (1986).
68. Hurst, L., *The Logic Processing of Digital Signals*, Crane Russak and Company Inc. (New York, NY, 1978).

69. "Basic ternary logic circuits," in *Circuits and Systems Based on Delta Modulation. Signals and Communication Technology*, Springer (Berlin, Heidelberg), pp. 41–50 (2005).
70. A.A. El-Slehdar, A.H. Fouad, A.G. Radwan, "Memristor-based balanced ternary adder", *25th International Conference on Microelectronics (ICM)*, pp. 1–4, 2013.
71. P.C. Balla, A. Antoniou, *IEEE J. Solid-State Circuits* 19(5), 739–749 (1984).
72. A. Heungand, H.T. Mouftah, *IEEE J. Solid-State Circuits* 20(2), 609–616 (1985).
73. M. Khalid, J. Singh, *Analog Integr. Circuits Signal Process.* 87(3), 399–406 (2016). DOI: 10.1007/s10470-016-0733-1.
74. K. Smagulova, O. Krestinskaya, A.P. James, *Analog Integr. Circuits Signal Process.* 95, 467 (2018).
75. G.S. Snider, *Nanotechnology* 18, 365202 (2007).
76. W. Manzoor, F. Bashir, F.A. Khanday, "Programmable threshold comparator using high frequency operational transconductance amplifier (OTA) based memristor", *3rd International Conference on Communication and Electronics Systems (ICCES 2018)*, 15–16 October 2018, PPG Institute of Technology, Coimbatore, India, 449–453, 2018.
77. T.C. Chang, K.C. Chang, T.M. Tsai, T.J. Chu, S.M. Sze, *Mater. Today* 19(5), 254–264 (2016).
78. Y.T. Su, H.W. Liu, P.H. Chen, T.C. Chang, T.M. Tsai, T.J. Chu, C.H. Pan, C.H. Wu, C.C. Yang, M.C. Wang, S. Zhang, H. Wang, S.M. Sze, *IEEE J. Solid-State Circuits* 6, 341–345 (2018).
79. S.X. Chen, S.P. Chang, S.J. Chang, W.K. Hsieh, C.H. Lin, *ECS J. Solid State Sci. Technol.* 7(7), Q3183–Q3188 (2018).
80. A. Prakash, D. Deleruyelle, J. Song, M. Bocquet, H. Hwang, *Appl. Phys. Lett.* 106(23), 233104 (2015).
81. G. Ma, X. Tang, H. Zhang, Z. Zhong, J. Li, H. Su, *Microelectron. Eng.* 139, 43–47 (2015).
82. F.M. Simanjuntak, D. Panda, K.H. Wei, T.Y. Tseng, *Nanoscale Res. Lett.* 11(1), 368 (2016).
83. S.X. Chen, S.P. Chang, W.K. Hsieh, S.J. Chang, C.C. Lin, *RSC Adv.* 8(32), 17622–17628 (2018).
84. M.K. Yang, J.W. Park, T.K. Ko, J.K. Lee, *Appl. Phys. Lett.* 95(4), 042105 (2009).
85. F.C. Chiu, W.C. Shih, J.J. Feng, *J. Appl. Phys.* 111, 9 (2012).
86. W. Banerjee, X. Xu, H. Liu, H. Lv, Q. Liu, H. Sun, S. Long, M. Liu, *IEEE Trans. Electron Device Lett.* 36(4), 333–335 (2015).
87. M.C. Wu, W.Y. Jang, C.H. Lin, T.Y. Tseng, *Semicond. Sci. Technol.* 27(6), 065010 (2012).
88. H.S.P. Wong, H.Y. Lee, S. Yu, Y.S. Chen, Y. Wu, P.S. Chen, B. Lee, F.T. Chen, M.J. Tsai, *Proc. IEEE* 100(6), 1951–1970 (2012).
89. L. Zhao, Z. Jiang, H. Y. Chen, J. Sohn, K. Okabe, B. Magyari-Kope, H.S.P. Wong, Y. Nishi, "Ultrathin (~2nm) HfOx as the fundamental resistive switching element: Thickness scaling limit, stack engineering and 3D integration", *2014 IEEE International Electron Devices Meeting (IEDM)*, 6.6.1–6.6.4, 2014.
90. X. Yang, S. Long, K. Zhang, X. Liu, G. Wang, X. Lian, Q. Liu, H. Lv, M. Wang, H. Xie, H. Sun, P. Sun, J. Sune, M. Liu, *J. Phys. D Appl. Phys.* 46, 24 (2013).
91. G. Wang, S. Long, M. Zhang, Y. Li, X. Xu, H. Liu, M. Wang, P. Sun, H. Sun, Q. Liu, H. Lu, B. Yang, M. Liu, *Sci. China Technol. Sci.* 57(12), 2295–2304 (2014).
92. R. Waser, *Microelectron. Eng.* 86(7–9), 1925–1928 (2009).
93. M.N. Kozicki, H.J. Barnaby, *Semicond. Sci. Technol.* 31(11), 113001 (2016).
94. L. Goux, I. Valov, *Phys. Status Solidi (A)* 213(2), 274–288 (2016).
95. Y. Huang, Z. Shen, Y. Wu, X. Wang, S. Zhang, X. Shia, H. Zeng, *RSC Adv.* 6(22), 17867–17872 (2016).

96. "2015, ITRS-ERD meeting", *The International Technology Roadmap for Semiconductors*, 2015.
97. J.W. Jang, B. Attarimashalkoubeh, A. Prakash, H. Hwang, Y.H. Jeong, *IEEE Trans. Electron Devices* 63(6), 2610–2613 (2016).
98. M. Lanza, H.-S. P. Wong, E. Pop, D. Ielmini, D. Strukov, B. C. Regan, L. Larcher, M.A. Villena, J.J. Yang, L. Goux, A. Belmonte, Y. Yang, F.M. Puglisi, J. Kang, B.M. Kope, E. Yalon, A. Kenyon, M. Buckwell, A. Mehonic, A. Shluger, H. Li, T.H. Hou, B. Hudec, D. Akinwande, R. Ge, S. Ambrogio, J.B. Roldan, E. Miranda, J. Sune, K.L. Pey, X. Wu, N. Raghavan, E. Wu, W.D. Lu, G. Navarro, W. Zhang, H. Wu, R. Li, A. Holleitner, U. Wurstbauer, M.C. Lemme, M. Liu, S. Long, Q. Liu, H. Lv, A. Padovani, P. Pavan, I. Valov, X. Jing, T. Han, K. Zhu, S. Chen, F. Hui, Y. Shi, *Adv. Electron. Mater.* 5(1) 1800143 (2018)
99. G.G. Cordero, F.J. Molinos, J.B.R.B. Gonzalez, F. Campabadal, *J. Vac. Sci. Technol. B Nanotechnol. Microelectron. Mater. Process. Meas. Phenom.* 35(1), 01A110 (2017).
100. Q. Wu, W. Banerjee, J. Cao, Z. Ji, L. Li, M. Liu, *Appl. Phys. Lett.* 113, 2 (2018).
101. W. Chen, W. Lu, B. Long, Y. Li, D. Gilmer, G. Bersuker, S. Bhunia, R. Jha, *Semicond. Sci. Technol.* 30(7), 075002 (2015).
102. B. Butcher, G. Bersuker, K.G. Young-Fisher, D.C. Gilmer, A. Kalantarian, Y. Nishi, R. Geer, P.D. Kirsch, R. Jammy, "Hot forming to improve memory window and uniformity of low-power HfOx-based RRAMs", *2012 4th IEEE International Memory Workshop*, 1–4, May 2012.
103. L. Zhao, H. Y. Chen, S. C. Wu, Z. Jiang, S. Yu, T.H. Hou, H.S.P. Wong, Y. Nishi, *Nanoscale* 6(11), 5698–5702 (2014).
104. C. Wang, H. Wu, B. Gao, T. Zhang, Y. Yang, H. Qian, *Microelectron. Eng.* 187, 121–133 (2018).
105. H. Abdalla, M.D. Pickett, "SPICE modeling of memristors", in *IEEE International Symposium on Circuits and Systems*, Piscataway, 2011.
106. C. Yakopcic, T.M. Taha, G. Subramanyam, R.E. Pino, *IEEE Trans. Comput. Aided Design Integ. Circuits Syst.* 32(8), 1201–1204 (2013).
107. S. Kvatinsky, E.G. Friedman, A. Kolodny, S. Member, U.C. Weiser, *IEEE Trans. Circuits Syst.* 60(1), 211–221 (2013).
108. S. Kvatinsky, M. Ramadan, E.G. Friedman, A. Kolodny, *IEEE Trans. Circuits Syst.* 62(8), 786–790 (2015).
109. Synopsys Inc., *HSPICE Reference Manual*, Synopsys Inc., (Mountain View, CA, 2009).
110. S. Kim, S.-J. Kim, K.M. Kim, S.R. Lee, M. Chang, E. Cho, *Sci. Rep.* 3, 1680 (2013).
111. P. Huang, X.Y. Liu, B. Chen, H.T. Li, Y.J. Wang, Y.X. Deng, *IEEE Trans. Electron Devices* 60, 4090–4097 (2013).
112. M. Bocquet, D. Deleruyelle, H. Aziza, C. Muller, J.M. Portal, T. Cabout, *IEEE Trans. Electron Devices* 61, 674–681 (2014).
113. D. Berco, T.Y. Tseng, *AIP Adv.* 6 (2016).
114. G. González-Cordero, J. Roldan, F. Jiménez-Molinos, J. Suñé, M. Liu, *Semicond. Sci. Technol.* 31, 115013 (2016).
115. S.C. Chae, J.S. Lee, S. Kim, S.B. Lee, S.H. Chang, C. Liu, *Adv. Mater.* 20, 1154–1159 (2008).
116. U. Russo, D. Ielmini, C. Cagli, A.L. Lacaita, *IEEE Trans. Electron Devices* 56, 186–192 (2009).
117. F. Corinto, S. Member, A. Ascoli, *IEEE Trans. Circuits Syst.* 59, 2713–2726 (2012).
118. F. Zahoor, T.Z.A.B. Zulkifli, F.A. Khanday, A.A. Fida, "1T1R Array Design with CNTFET as Access Device", *2019 IEEE Student Conference on Research and Development (SCOReD)*, October 15–17, 2019, Seri Iskandar, Perak, Malaysia), pp. 280–283, 2019.

119. G.G. Cordero, J.B. Roldan, F.J. Molinos, "Simulation of RRAM memory circuits, a Verilog-A compact modeling approach", *2016 Conference on Design of Circuits and Integrated Systems (DCIS)*, pp. 1–6, 2016.
120. M.C. Wu, W.Y. Jang, C.H. Lin, T.Y. Tseng, *Semicond. Sci. Technol.* 27, 065010 (2012).
121. U. Russo, D. Kamalnathan, D. Ielmini, A.L. Lacaita, M.N. Kozicki, *IEEE Trans. Electron Devices* 56(5), 1040–1047 (2009).
122. H. Li, B. Gao, Z. Chen, Y. Zhao, P. Huang, H. Ye, L. Liu, X. Liu, J.Kang, *Sci. Rep.* 5, 13330 (2015).
123. J.J. Yang, D.B. Strukov, D.R. Stewart, *Nat. Nanotechnol.* 8(1), 13–24 (2013).
124. S. Gao, F. Zeng, M. Wang, G. Wang, C. Song, F. Pan, *Sci. Rep.* 5(1), 15467 (2015).
125. S. Balatti, S. Ambrogio, D. Ielmini, *IEEE Trans. Electron Devices* 62(6), 1831–1838 (2015).
126. M. Prezioso, F. Merrikh-Bayat, B.D. Hoskins, G.C. Adam, K.K. Likharev, D.B. Strukov, *Nature* 521(7750), 61–64 (2015).
127. H. Wu, X.H. Wang, B. Gao, N. Deng, Z. Lu, B. Haukness, G. Bronner, H. Qian, *Proc. IEEE* 105(9), 1770–1789 (2017).

Index

C

D

E

F

G

H

I

N

O

P

Q

R

S

T

U

V

W

X

Y

Z